STRATEGIES
OF
PROBLEM SOLVING

Second Edition

Maria Nogin

Department of Mathematics

College of Science and Mathematics

California State University, Fresno

2014

Acknowlegdements

I am very thankful to my parents for introducing me to the beautiful world of mathematics problem solving. Since very young age, they constantly asked me intriguing questions answering which required logical thinking, counting, experimenting, and noticing patterns. They bravely gave me very hard problems, but encouraged me when I got stuck. Some problems took me up to two years to solve. But they never just gave answers to me. They believed in my ability to solve those problems on my own.

I also would like to thank my colleagues in the Mathematics Department at CSU Fresno, especially Agnes Tuska and Larry Cusick, for numerious valuable discussions about interesting problems and solutions as well as teaching and evaluating problem solving. Also thank you to my Fall 2013 Math 145 (Problem Solving) class for their careful reading and catching many errors in the first edition.

Finally, I am thankful to both my parents and my husband for encouraging me to write this book.

To my daughters Katya and Misha.

I hope you will grow to love mathematics
and enjoy problem solving as much as I do.

Table of Contents

Chapter 1

Introduction

Solving mathematical problems is both a science and an art. It is a science because we need to learn some basic concepts and skills, and use proper terminology when explaining our solution to other people. It is also an art because very often we need to be creative. There are infinitely many types of math problems. While it is important to learn some basic principles of problem solving, it is impossible to learn how to solve every problem in the world. Just like it is impossible to learn how to construct every possible model using Lego Blocks. However, after you've constructed a few following the directions in your Lego set, you can create your own. When you finish school and move on to solving real world problems, you may need some initial training in your field, but no training can ever give you detailed instructions on what to do in every possible situation. You'll have to think, make decisions, try things out, and learn from your successes and mistakes. This is what we are going to do in this book: learn some basics and then explore on our own and learn from our experience.

Below are some problems. Let's try solving them. Solutions to these problems will illustrate some things mentioned above as well as will introduce a few concepts and principles that will be discussed in later chapters.

1. Eleven children contributed money to buy a present for their classmate. The total amount of money collected was $30.00. Prove that at least one child gave at least $2.73.

2. (a) Prove that any two-digit number is divisible by 3 if and only if the sum of its digits is divisible by 3.

   (b) Prove that any natural number is divisible by 3 if and only if the sum of its digits is divisible by 3.

3. Is it true or false that for any natural number n, the number $n^2 + n + 41$ is prime?

4. In a 4×4 table six cells are marked by a star and all others are blank. Show that it is possible to cross out 2 columns and 2 rows so that the remaining cells are blank.

5. Is it true or false that for any natural number n, the number $n^3 + 2n$ is divisible by 3?

6. Sketch the graph of $f(x) = |x + 2| + |2x - 5|$.

7. Konigsberg is a city which was the capital of East Prussia but now is known as Kaliningrad in Russia. The city is built around the River Pregel where it joins another river. An island named Kniephof is in the middle of where the two rivers join. There are

seven bridges that join the different parts of the city on both sides of the rivers and the island.

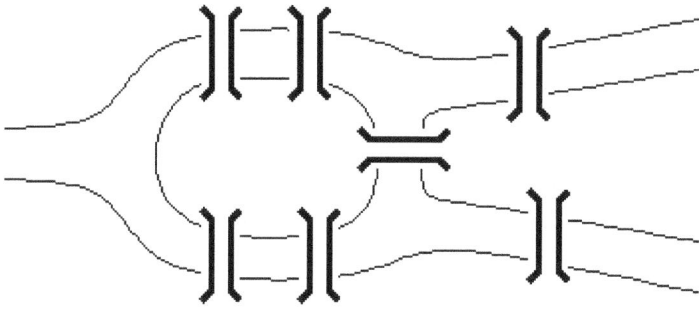

People tried to find a way to walk all seven bridges without crossing a bridge twice, but no one could find a way to do it. The problem came to the attention of a Swiss mathematician named Leonhard Euler. In 1735, Euler presented the solution to the problem before the Russian Academy.

Now, you too try to solve this problem. If such a tour exists, find it. If not, explain why not.

8. (a) Is it possible for a chess knight to start at the upper left corner and go through every square on the 8 × 8 chessboard exactly once? (A knight's move is 2 squares up, down, or to the right or left, and 1 square in a perpendicular direction. All allowed moves from a certain square are shown below.)

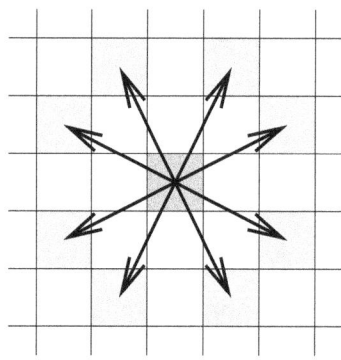

   (b) Is it possible for a knight to start at the upper left corner, go through every square on the 8 × 8 chessboard exactly once, and come back to the starting point?

As said above, learning to solve problems is in part difficult because problems can be very different. However, there are a few basic principles that are good to know. There are a few approaches and methods that can be useful. In this book, we'll study some of them. After you study the material of this book you should be able to solve many problems pretty easily.

While using intuition and working out a few examples may help us find an idea, it is also important to write rigorous proofs. Since our intuition is not always correct, we need to justify each step in a solution. We will therefore try to avoid words such as 'obviously'.

In each chapter, we provide basic definitions and facts to get you started. We do not prove most of the facts given in this book, since our main goal is to learn how to solve problems, i.e. use these facts. You will probably prove most, if not all, the facts given in this book in courses such as Calculus, Discrete Mathematics, Abstract Algebra, and Number Theory. Sometimes the idea of a proof of a theorem can be used for solving many problems. In such cases we provide the proof.

Chapter 2

Introduction to Logic

In this chapter we will introduce basic logic terminology and notations. It will be useful when we discuss types of proofs (see chapter 3).

Definition 2.1. A **proposition** is a declarative sentence that is either true or false.

For example, "3 plus 2 is 5" is a true proposition, "3 times 2 is 7" is a false proposition, while "x minus 4 is 8" is not a proposition because the value of x has not been defined, and "is 3 plus 3 equal 6?" is not a proposition because it is an interrogative, not declarative, sentence.

Definition 2.2. Let p and q be propositions. Then:

- The **negation** of p, denoted by $\neg p$, is the proposition "not p". It is true if and only if p is false.

- The **conjunction** of p and q, denoted by $p \wedge q$, is the proposition "p and q". It is true if and only if both p and q are true.

- The **disjunction** of p and q, denoted by $p \vee q$, is the proposition "p or q". It is true if and only if at least one of p and q is true. Note that if both p and q are true, then $p \vee q$ is true, so "or" is not exclusive.

- The **exclusive or** of p and q, denoted by $p \oplus q$, is the proposition "either p or q but not both". It is true if and only if exactly one of p and q is true.

- The **implication** of p and q, denoted by $p \rightarrow q$, is the proposition "if p then q". It is false when p is true and q is false, and true otherwise.

- The **biconditional** of p and q, denoted by $p \leftrightarrow q$, is the proposition "p if and only if q". It is true when p and q have the same truth values and is false otherwise.

Below is the so-called truth table that shows the truth values of the compound propositions defined above depending on the truth values of p and q.

p	q	$\neg p$	$p \wedge q$	$p \vee q$	$p \oplus q$	$p \rightarrow q$	$p \leftrightarrow q$
T	T	F	T	T	F	T	T
T	F	F	F	T	T	F	F
F	T	T	F	T	T	T	F
F	F	T	F	F	F	T	T

Notice that $p \rightarrow q$ is false if and only if p is true and q is false. We will need this observation in chapter 3.

- Symbols $\neg$, $\wedge$, $\vee$, $\oplus$, $\rightarrow$, and $\leftrightarrow$ are called **logical connectives**.

- A compound proposition that is always true, no matter what the truth values of the propositions that occur in it are, is called a **tautology**.

  For example, $p \vee \neg p$ is a tautology.

- A compound proposition that is always false is called a **contradiction**.

  For example, $p \wedge \neg p$ is a contradiction.

- Propositions p and q are called **logically equivalent** if $p \leftrightarrow q$ is a tautology. The notation $p \Leftrightarrow q$ denotes that p and q are logically equivalent. Note that p and q are logically equivalent if and only if they always have the same truth values.

Example 2.3. Show that $\neg(p \vee q)$ and $(\neg p) \wedge (\neg q)$ are logically equivalent.

Solution. *We construct the truth table:*

p	q	$p \vee q$	$\neg(p \vee q)$	$\neg p$	$\neg q$	$(\neg p) \wedge (\neg q)$
T	T	T	F	F	F	F
T	F	T	F	F	T	F
F	T	T	F	T	F	F
F	F	F	T	T	T	T

We see that the truth values of $\neg(p \vee q)$ and $(\neg p) \wedge (\neg q)$ are always the same, therefore these propositions are logically equivalent.

Definition 2.4. A statement $P(x)$ that depends on the value of a variable (x in this case) is called a **propositional function**. Once a value has been assigned to the variable x, the statement $P(x)$ becomes a proposition and has a truth value.

For example, if $P(x)$ is the statement "$x > 3$", then $P(4)$ is true and $P(2)$ is false.

Definition 2.5. Let $P(x)$ be a propositional function. Then

- $\forall x P(x)$ means "for every x, $P(x)$ is true".

- $\exists x P(x)$ means "there exists a value of x for which $P(x)$ is true".

- $\exists! x P(x)$ means "there exists a unique value of x for which $P(x)$ is true".

Symbols $\forall$ and $\exists$ are called **quantifiers**. Namely, $\forall$ is called a **universal quantifier**, and $\exists$ is called an **existential quantifier**.

When interpreting expressions $\forall x P(x)$, $\exists x P(x)$, $\exists! x P(x)$, we need to specify a set S of all possible choices of x. Such a set is called the **domain of discourse**. Unless the domain of discourse has already been specified or is clear from context, we can write $\forall x \in S\ P(x)$, etc. to make it explicit. For example, "the square of every integer x is nonnegative" can be written as $\forall x \in \mathbb{Z}\ x^2 \geq 0$.

Propositional functions can be functions of two or more variables, and then we can use two or more quantifiers with them. It is important to realize that the order of quantifiers makes a difference. For example, below we will use the propositional function $F(x, y)$ which means that x and y are friends (the domain of this function can be a set of people). Then e.g. $\forall x \exists y F(x, y)$ means that everybody has at least one friend, while $\exists y \forall x F(x, y)$ means that there is a person who is friends with everybody.

Propositions with negations can always be written so that no negation is outside a quantifier or an expression involving logical connectives, for example:

- $\neg(p \wedge q) \Leftrightarrow \neg p \vee \neg q$
- $\neg(p \vee q) \Leftrightarrow \neg p \wedge \neg q$
- $\neg \forall x P(x) \Leftrightarrow \exists x \neg P(x)$
- $\neg \exists x P(x) \Leftrightarrow \forall x \neg P(x)$

Problems

1. Show that the following propositions are logically equivalent.

   (a) $p \to q$ and $\neg q \to \neg p$
   (b) $p \to q$ and $\neg p \vee q$
   (c) $\neg(p \wedge q)$ and $\neg p \vee \neg q$
   (d) $p \vee (q \wedge r)$ and $(p \vee q) \wedge (p \vee r)$

2. Which of the following sentences are statements? For those that are, indicate the truth value.

   (a) Five plus eight is thirteen.
   (b) Five minus eight is three.
   (c) Two times x is 6.
   (d) The number 2n+6 is an even integer.
   (e) There are 200 elephants in the San Diego Wild Animal Park.
   (f) I have solved all problems in chapter 1.
   (g) Did you do your homework today?

3. Translate the statement

$$\forall x (C(x) \vee \exists y (C(y) \wedge F(x, y)))$$

   into English, where $C(x)$ is "x has a computer", $F(x, y)$ is "x and y are friends", and the domain of discourse is the set of all students at your university.

4. Let $F(x, y)$ be statement "x can fool y". Use quantifiers to express each of the following statements:

   (a) Everybody can fool Amy.
   (b) Mike can fool everybody.
   (c) Everybody can fool somebody.
   (d) There is no one who can fool everybody.
   (e) Everyone can be fooled by somebody.
   (f) No one can fool both Kate and Jerry.
   (g) Tim can fool exactly two people.
   (h) There is exactly one person whom everybody can fool.
   (i) No one can fool himself or herself.

5. Let $P(x)$ denote the propositional function "$x = -5$" and let $Q(x)$ denote the propositional function "$x^2 = 25$" and let the domain of discourse be the set of real numbers. Determine the truth values of the following propositions:

   (a) $P(4)$

   (b) $P(4) \rightarrow Q(4)$

   (c) $\exists x \neg P(x)$

   (d) $\forall x (P(x) \vee Q(x))$

   (e) $\exists x (P(x) \wedge Q(x))$

   (f) $\forall x (P(x) \rightarrow Q(x))$

   (g) $\exists x (P(x) \rightarrow Q(x))$

   (h) $\forall x (P(x) \leftrightarrow Q(x))$

6. Let $P(x)$ denote the propositional function "$(x < 3) \vee (x > 5)$" and let the domain of discourse be the set of real numbers. Determine the truth values of the following propositions.

   (a) $P(2)$

   (b) $P(4)$

   (c) $P(2) \wedge P(4)$

   (d) $\forall x P(x)$

   (e) $\exists x P(x)$

   (f) $\exists! x P(x)$

   (g) $\forall x (P(x) \vee P(-x))$

7. Let $Q(x, y)$ denote "$x + y = 0$" and let the domain of discourse be the set of real numbers. What are the truth values of the statements $\forall x \exists y Q(x, y)$ and $\exists y \forall x Q(x, y)$?

8. Rewrite each of the following statements so that negations appear only immediately before propositional functions.

   (a) $\neg \forall x \forall y P(x, y)$

   (b) $\neg \forall y \exists x P(x, y)$

   (c) $\neg \forall y \forall x (P(x, y) \vee Q(x, y))$

   (d) $\neg (\exists x \exists y \neg P(x, y) \wedge \forall x \forall y Q(x, y))$

   (e) $\neg \forall x (\exists y \forall z P(x, y, z) \wedge \exists z \forall y P(x, y, z))$

   (f) $\neg \exists! x P(x)$

9. Let $P(x, y)$ denote the proposition "$x < y$" and let the domain of discourse be the set of real numbers. Determine the truth values of the following propositions.

   (a) $\exists x \exists y P(x, y)$,

   (b) $\forall x \exists y P(x, y)$,

   (c) $\exists x \forall y P(x, y)$,

   (d) $\forall x \forall y P(x, y)$,

   (e) $\forall x P(-x, x)$.

10. Let $R(x, y)$ be the statement "$x + y = x - y$" and let the domain of discourse be the set of integers. Find the truth values of the following statements. Explain.

  (a) $R(2, 0)$

  (b) $\forall y R(1, y)$

  (c) $\forall x \exists y R(x, y)$

  (d) $\forall y \exists x R(x, y)$

  (e) $\exists y \forall x R(x, y)$

11. Which of the following compound propositions are logically equivalent, i.e. have the same truth values of any propositional functions $P(x)$ and $Q(x)$? If propositions are logically equivalent, explain why. If not, give an example of propositional functions $P(x)$ and $Q(x)$ for which one of the propositions is true and the other one is false.

  (a) $\forall x (\neg P(x))$ and $\neg(\forall x P(x))$

  (b) $\forall x (P(x) \vee Q(x))$ and $(\forall x P(x)) \vee (\forall x Q(x))$

  (c) $\forall x (P(x) \wedge Q(x))$ and $(\forall x P(x)) \wedge (\forall x Q(x))$

  (d) $\forall x (P(x) \rightarrow Q(x))$ and $(\forall x P(x)) \rightarrow (\forall x Q(x))$

  (e) $\forall x (P(x) \leftrightarrow Q(x))$ and $(\forall x P(x)) \leftrightarrow (\forall x Q(x))$

  (f) $\exists x (\neg P(x))$ and $\neg(\exists x P(x))$

  (g) $\exists x (P(x) \vee Q(x))$ and $(\exists x P(x)) \vee (\exists x Q(x))$

  (h) $\exists x (P(x) \wedge Q(x))$ and $(\exists x P(x)) \wedge (\exists x Q(x))$

  (i) $\exists x (P(x) \rightarrow Q(x))$ and $(\exists x P(x)) \rightarrow (\exists x Q(x))$

  (j) $\exists x (P(x) \leftrightarrow Q(x))$ and $(\exists x P(x)) \leftrightarrow (\exists x Q(x))$

12. Express the definition of the limit $\lim\limits_{x \to a} f(x) = L$ using quantifiers.

13. Express the definition of a convergent sequence $a_1, a_2, \ldots$ using quantifiers.

14. In one country there are two cities, A and B, that are only a few miles apart, and whose residents often visit each other. All residents of city A always say the truth, while all residents of city B always lie. A stranger is passing through one of these cities, but he doesn't know which one. How could he, by asking the first man he sees only one question, determine which city he is passing through?

Chapter 3

Types of proofs

In this chapter we summarize basic types of proofs, and then give a few examples to illustrate them.

Suppose we want to prove a proposition p.

- a **direct** proof just shows that p holds;

- a proof **by contradiction** assumes that p is false and derives a contradiction. The contradiction is usually of the form $r \wedge \neg r$ for some proposition r.

If we want to prove an implication "if p, then q", then any of the following types of proofs may be used:

- a **direct** proof just shows how q follows from p;

- a proof **by contradiction** assumes that $p \rightarrow q$ is false, i.e. p is true and q is false, and derives a contradiction;

- a proof **by contrapositive** shows that $\neg q$ implies $\neg p$.

A proof of a statement of the form "$\exists x P(x)$" can be

- **constructive** - when we provide (construct) such an x explicitly;

- **existential**, or **nonconstructive** - when we show the existence of such an x without actually constructing it.

To prove a statement of the form "$\forall x P(x)$" where the domain of discourse is a subset of integer numbers, it is often (but not always!) a good idea to use Mathematical Induction (see chapter 4).

To prove a statement of the form "$p \leftrightarrow q$", we can either

- prove $p \rightarrow q$ and $q \rightarrow p$ separately, or

- have each step of our proof of the form "if and only if".

To **disprove** a statement means to show that it is false. To disprove a statement of the form $\forall x P(x)$ it is sufficient to show that there exists at least one **counterexample**, that is, there exists at least one case when the statement does not hold.

Below are some examples of various types of proofs listed above.

Example 3.1. Prove that every odd integer is the difference of two perfect squares.

Direct proof: Every odd integer has the form $2n + 1$ for some integer n. Observe that $2n + 1 = (n + 1)^2 - n^2$.

Example 3.2. Prove that $\sqrt{2}$ is irrational.

Proof by contradiction: Suppose $\sqrt{2}$ is rational. Then there exists an irreducible fraction $\dfrac{p}{q} = \sqrt{2}$. (Irreducible means that the greatest common divisor of p and q is 1.) Then $\dfrac{p^2}{q^2} = 2$, thus $p^2 = 2q^2$. If follows that p^2 is even, so p is even. Let $p = 2m$, where m is an integer, then $p^2 = 4m^2$. We have $4m^2 = 2q^2$, or $2m^2 = q^2$. Now we see that q^2 is even, therefore q is even. We get a contradiction because we have that on one hand, p and q have the greatest common divisor 1, but on the other hand p and q are both even.

Example 3.3. Prove that if a and b are integers and ab is even, then either a or b is even (or both).

Proof by contrapositive: Suppose that neither a nor b is even, and we will prove that ab is not even. That is, we suppose that both a and b are odd, and we will prove that ab is odd. Any odd numbers a and b can be written in the form $a = 2n + 1$ and $b = 2m + 1$ for some integers n and m. Then we have $ab = (2n + 1)(2m + 1) = 4nm + 2n + 2m + 1 = 2(2nm + n + m) + 1$ is an odd number.

Example 3.4. Prove that for every positive integer n, there exist n consecutive composite numbers.

Constructive proof: We claim that $(n + 1)! + 2$, $(n + 1)! + 3$, ... , $(n + 1)! + (n + 1)$ are all composite. Indeed, $(n+1)!$ is divisible by 2, by 3, ... , and by $n+1$. Therefore $(n+1)!+2$ is divisible by 2, $(n + 1)! + 3$ is divisible by 3, ... , $(n + 1)! + (n + 1)$ is divisible by $n + 1$.

Example 3.5. Prove that $x^3 + x - 1 = 0$ has a real root.

Nonconstructive proof: Let $f(x) = x^3 + x - 1$. Then $f(-1) = -3 < 0$ and $f(1) = 1 > 0$. Since $f(x)$ is a polynomial, it is continuous. By the Intermediate Value Theorem, there exists c between -1 and 1 such that $f(c) = 0$.

Example 3.6. Prove or disprove that every odd integer is prime.

Counterexample: 9 is odd but not prime. Thus the statement is false.

Problems

1. Prove that if n is an integer and $3n + 5$ is odd, then n is even. Is your proof direct, by contradiction, or by contrapositive?

2. Prove that an integer a is even if and only if a^2 is even. Did you prove the two implications separately or simultaneously?

3. Prove or disprove that $2^n + 1$ is prime for all nonnegative integers n.

4. Prove that for any integer n there is a prime number greater than n. Is your proof constructive?

5. Every odd number is either of the form $4n + 1$ (if it has remainder 1 when divided by 4) or of the form $4n + 3$ (if it has remainder 3) where n is an integer. Prove that if an odd number is a perfect square, then it has the form $4n + 1$. What type of proof did you use? State the converse. Prove or disprove the converse.

6. Prove or disprove that if a and b are rational numbers, then a^b is also rational.

7. Prove that the equation $x^{101} + x^{51} + x + 1 = 0$ has exactly one real solution. Split this into two statements:

    (a) the equation has at least one solution. Is your proof constructive or nonconstructive?

    (b) the equation can not have two distinct roots. Is your proof direct, by contradiction, or by contrapositive?

8. Prove that if the sum of two numbers is irrational then at least one of the numbers is irrational. Is your proof direct, by contradiction, or by contrapositive? State the converse. Prove or disprove the converse.

9. Prove that the equation $4\sin^2 x = 1$ has a real solution. Is your proof constructive?

10. Prove that the equation $x + \sin x = 1$ has a real solution. Is your proof constructive?

11. Prove that the equation $x^2 + x + 1 = 0$ has no rational solutions. Is your proof direct, by contradiction, or by contrapositive?

12. Prove that 0 is a root of the equation $a_n x^n + \ldots a_1 x + a_0 = 0$ if and only if the free term $a_0 = 0$. Did you prove the two implications separately or simultaneously?

13. Prove that if a positive integer is divisible by 8 then it is the difference of two perfect squares. Is your proof direct, by contradiction, or by contrapositive? Is it constructive or nonconstructive?

14. Prove or disprove that if a and b are irrational numbers, then a^b is also irrational.

15. Prove that for any integers n and m, if $nm + 2n + 2m$ is odd then both n and m are odd. Is your proof direct, by contradiction, or by contrapositive?

Chapter 4

Principle of Mathematical Induction

Theorem 4.1. *(Principle of Mathematical Induction) Let S_n be a statement about a positive integer n. Suppose that*

1. *S_1 is true,*

2. *If $k \geq 1$ and S_k is true, then S_{k+1} is true.*

Then S_n is true for all positive integers n.

Note. Conditions 1 and 2 in the above theorem are called the basis step and inductive step respectively.

This principle is easy to understand using the following example: suppose we know how to get to the first floor of a building (e.g. we know where an entrance is), and we also know how to get from any floor to the next one (e.g. we know where an elevator or a staircase is). Then we'll be able to get to any floor in this building. Namely, we'll be able get to the first floor, and then from the first to the second, and then from the second to the third, and so on. The same is true for any statement. If we can check that S_1 is true, then the second condition in theorem 4.1 ensures that S_2 follows from S_1, S_3 follows from S_2, and so on. Thus S_n is true for any natural number n.

Mathematical Induction is used in all areas of mathematics. It can be used to prove summation formulas such as in the next example, various number theory, algebraic, and geometric statements.

Example 4.2. Prove that for any natural number n,

$$1 + 2 + 3 + \ldots + n = \frac{n(n+1)}{2}.$$

Proof. We will prove this identity using Mathematical Induction.

Basis step: if $n = 1$, the formula says that $1 = \dfrac{1 \cdot (1+1)}{2}$ which is true.

Inductive step: suppose the formula holds for $n = k$, i.e. that

$$1 + 2 + 3 + \ldots + k = \frac{k(k+1)}{2} \tag{4.1}$$

17

is true. We have to show that the formula holds for $n = k+1$, i.e. that

$$1 + 2 + 3 + \ldots + (k+1) = \frac{(k+1)((k+1)+1)}{2}$$

is true. Adding $k+1$ to both sides of (4.1) gives:

$$
\begin{aligned}
1 + 2 + 3 + \ldots + k + (k+1) &= \frac{k(k+1)}{2} + (k+1) \\[2mm]
&= \frac{k(k+1) + 2(k+1)}{2} \\[2mm]
&= \frac{(k+2)(k+1)}{2} \\[2mm]
&= \frac{((k+1)+1)(k+1)}{2}.
\end{aligned}
$$

$\square$

Note. For any specific value of n, it is easy to check that the identity holds. For example, for the first four natural numbers we have:

$$1 = \frac{1 \cdot (1+1)}{2}, \quad 1 + 2 = \frac{2 \cdot (2+1)}{2}, \quad 1 + 2 + 3 = \frac{3 \cdot (3+1)}{2}, \quad 1 + 2 + 3 + 4 = \frac{4 \cdot (4+1)}{2}.$$

However, remember that it is not sufficient to check *some* values of n. We had to prove the statement for *all* natural numbers n.

Remark. We might want to prove a statement S_n for all $n \geq 0$, or for all $n \geq 2$, etc., rather than for all $n \geq 1$. In this case, the basis step should check that the statement is valid for the smallest value of n, say, $n = 0$, or $n = 2$ in the above cases, and the inequality $k \geq 1$ in the inductive step should be modified accordingly ($k \geq 0$, or $k \geq 2$, etc.).

Sometimes to prove S_{k+1}, it is insufficient to assume S_k alone, but S_n for $n \leq k$ is needed. Then we use the so-called Strong Induction formulated below.

Theorem 4.3. *(Strong Mathematical Induction) Let S_n be a statement about a positive integer n. Suppose that*

 1. S_1 is true,

 2. If $k \geq 1$ and S_n is true for all $1 \leq n \leq k$, then S_{k+1} is true.

Then S_n is true for all positive integers n.

Remark. As above, we might want to start with 0 or 2 or something else rather than with 1.

Example 4.4. Prove that any integer $n \geq 2$ can be written in the form $n = 2a + 3b$ for some nonnegative integers a and b (we will say that n is a nonnegative linear combination of 2 and 3).

Proof. Basis step. If $n = 2$, we have $n = 2 \cdot 1 + 3 \cdot 0$.
Inductive step. Suppose that $k \geq 2$ and the statement holds for all $2 \leq n \leq k$. We want to prove it for $n = k+1$.
Case I. $k = 2$, so $k + 1 = 3$. Then $k + 1 = 3 = 2 \cdot 0 + 3 \cdot 1$.
Case II. $k \geq 3$, then $2 \leq k - 1 \leq k$, thus the statement holds for $n = k - 1$. We have $k - 1 = 2a + 3b$ for some nonnegative integers a and b. Then $k + 1 = k - 1 + 2 = 2a + 3b + 2 = 2(a+1) + 3b$, so $k + 1$ is a nonnegative linear combination of 2 and 3. $\square$

Remark. Notice that case I above simply checks that the statement holds for $n = 3$. In literature, this calculation is often moved to the basis step.

Problems

1. Prove that the following formulas hold for any natural n.

   (a) $1^2 + 2^2 + 3^2 + \ldots + n^2 = \dfrac{n(n+1)(2n+1)}{6}$

   (b) $1^3 + 2^3 + 3^3 + \ldots + n^3 = \left(\dfrac{n(n+1)}{2}\right)^2$

   (c) $1 \cdot 1! + 2 \cdot 2! + \ldots + n \cdot n! = (n+1)! - 1$

   (d) $1 \cdot 2 + 2 \cdot 3 + 3 \cdot 4 + \ldots + n(n+1) = \dfrac{n(n+1)(n+2)}{3}$

   (e) $1 + 3 + 5 + \ldots + (2n - 1) = n^2$

2. Prove that for any positive integer n, $n < 2^n$.

3. Prove that if q is a positive integer, then $3^{2^q} - 1$ is divisible by 2^{q+2}.

4. Suppose that $2n$ points are given in space, where $n \geq 2$. Altogether $n^2 + 1$ line segments are drawn between these points. Prove that there is at least one triangle (a set of three points which are joined pairwise by line segments).

5. Let $\{F_0, \ F_1, \ F_2, \ \ldots\}$ be the Fibonacci sequence defined by $F_0 = 0$, $F_1 = 1$, and $F_{n+1} = F_n + F_{n-1}$, $n \geq 1$. Prove the following identities.

   (a) $F_1 F_2 + F_2 F_3 + \ldots + F_{2n-1} F_{2n} = F_{2n}^2$

   (b) $F_1^2 + F_2^2 + \ldots + F_n^2 = F_n F_{n+1}$

   (c) $F_{n-1} F_{n+1} = F_n^2 + (-1)^n$

   (d) $\begin{bmatrix} 1 & 1 \\ 1 & 0 \end{bmatrix}^n = \begin{bmatrix} F_{n+1} & F_n \\ F_n & F_{n-1} \end{bmatrix}$

   (e) $F_{n-1}^2 + F_n^2 = F_{2n-1}$

6. There are n identical cars on a circular track. Among all of them, they have just enough gas for one car to complete a lap. Show that there is a car which can complete a lap by collecting gas from other cars on its way around.

7. Every road in Sikinia is one-way. Every pair of cities is connected by exactly one direct road. Show that there exists a city which can be reached from every other city either directly or via at most one other city.

8. Suppose that n lines are given in the plane. They divide the plane into regions. Show that it is possible to color the plane with two colors so that no regions with a common boundary line are colored the same way. Such a coloring is called a proper coloring.

9. Consider a few points in the plane and a few line segments connecting some of them so that (1) no two line segments intersect, and (2) each point is connected with at least two other points (so there are no isolated points and there are no "hanging" line segments). Such line segments divide the plane into several regions. Such a picture is called a map. Prove that a map can be properly colored with two colors if and only if each point is connected with an even number of other points. (See problem 8 for definition of a proper coloring)

10. Let α be any real number such that $\alpha + \dfrac{1}{\alpha} \in \mathbb{Z}$. Prove that $\alpha^n + \dfrac{1}{\alpha^n} \in \mathbb{Z}$ for any $n \in \mathbb{N}$.

11. Prove that $1 < \dfrac{1}{n+1} + \dfrac{1}{n+2} + \ldots + \dfrac{1}{3n+1} < 2$.

12. Let n be any natural number. Consider all nonempty subsets of the set $\{1, 2, \ldots, n\}$, which do not contain any neighboring elements. Prove that the sum of the squares of the products of all numbers in these subsets is $(n+1)! - 1$. (For example, if $n = 3$, then such subsets of $\{1, 2, 3\}$ are $\{1\}$, $\{2\}$, $\{3\}$, and $\{1, 3\}$, and $1^2 + 2^2 + 3^2 + (1 \cdot 3)^2 = 23 = 4! - 1$.)

13. Prove that the determinant of the $n \times n$ matrix M_n with entries

$$m_{ij} = \begin{cases} 5 \text{ if } i = j \\ 1 \text{ if } |i - j| = 1 \\ 0 \text{ otherwise} \end{cases}$$

is equal to $\dfrac{1}{3}(4^{n+1} - 1)$.

14. Find the determinant of the $n \times n$ matrix A_n with entries

$$a_{ij} = \begin{cases} 2 \text{ if } i = j \\ 1 \text{ if } |i - j| = 1 \\ 0 \text{ otherwise} \end{cases} .$$

Hint: calculate the determinants of A_1, A_2, A_3, and A_4. Notice the pattern. Guess a formula for $\det A_n$, and then prove it by Mathematical Induction.

15. Prove that if any one square of a $2^n \times 2^n$ chessboard is removed, then the remaining board can be covered by L-trominoes, i.e. the figures consisting of 3 squares as shown below.

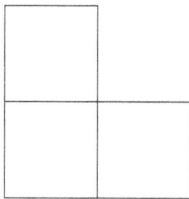

Chapter 5

Dirichlet's Box Principle

Theorem 5.1. *(Dirichlet's Box Principle) If $n + 1$ or more objects are put into n boxes, then at least one box contains more than one object.*

Dirichlet's Box Principle is often called the **Pigeonhole Principle** and is formulated as follows.

Suppose there are n pigeonholes in the tree, and there are at least $n + 1$ pigeons flying into these n holes. Then there is at least one hole containing more than one pigeon.

More formally and more generally, this principle can be formulated in the following way.

If the cardinality of a set S is bigger than the cardinality of a set T, and f is a function from S to T, then f is not one-to-one.

In the above theorem the function

$$S \xrightarrow{f} T$$

is

$$\{n + 1 \text{ objects}\} \longrightarrow \{n \text{ boxes}\}$$

or

$$\{n + 1 \text{ pigeons}\} \longrightarrow \{n \text{ pigeonholes}\}$$

Dirichlet's Box Principle is often used to prove statements involving remainders or divisibility. Recall that for any natural number n, there are n possible remainders upon division by n, namely, 0, 1, 2, ..., $n - 2$, and $n - 1$. If we are given more than n numbers, then by Dirichlet's Box Principle at least two of them have the same remainder. Note that this implies that the difference of these two numbers is divisible by n. Read more on remainders and divisibility in chapter 6.

Here is another, also very useful, principle.

Theorem 5.2. *If $n - 1$ or fewer objects are put into n boxes, then at least one box is empty.*

More formally,

If the cardinality of a set S is smaller than the cardinality of a set T, and f is a function from S to T, then f is not onto.

Below are generalizations of the above principles.

Theorem 5.3. *(Generalized Dirichlet's Box Principle) If $qn+1$ or more objects are put into n boxes, then at least one box contains more than q objects.*

Theorem 5.4. *If $n-k$ or fewer objects are put into n boxes, then at least k boxes are empty.*

Problems

1. Prove that among 13 persons, at least two were born in the same month.

2. Prove that among 50 persons, at least 5 were born in the same month.

3. Prove that among 120 distinct integers, there are two whose difference ends in 00.

4. The capital of Sikinia has 300,001 inhabitants, and it is known that none of them has more than 300,000 hairs on his or her head. Can you assert with certainty that there are two persons with the same number of hairs on their heads?

5. Seven points are given inside a regular hexagon whose sides have length 1. Prove that there are two among these seven points such that the distance between them is at most 1.

6. Suppose that five lattice points (i.e. points with integer coordinates) are given in the plane. Prove that we can choose two of these points such that the segment joining these two points passes through another lattice point.

7. Prove that from any 12 distinct two-digit numbers, we can select two with a two-digit difference of the form aa.

8. (a) Prove that from any 52 positive integers, we can select two such that their sum or difference is divisible by 100.

   (b) Is the above assertion also valid for 51 positive integers?

9. Three hundred points are given inside a cube with edge 7. Prove that we can place a small cube with edge 1 inside the big cube such that the interior of the small cube does not contain any of the given points.

10. Prove that if there are n persons present in a room, then among them there are two persons that have the same number of acquaintances.

11. Let a_1, a_2, and a_3 be integers. Show that the product $(a_1 - a_2)(a_1 - a_3)(a_2 - a_3)$ is even.

12. Let a_1, a_2, a_3, and a_4 be integers. Show that the product $\prod_{1 \le i < j \le 4} (a_i - a_j)$ is divisible by 12.

13. (a) Seven points are given inside a 9×12 rectangle. Prove that there are two of them such that the distance between them is less than 7.

    (b) Is the above assertion also valid for six points?

14. Twenty-six points are given inside a 20×15 rectangle. Prove that there are at least two points with distance less than or equal to 5.

15. Kevin is paid every other week on Friday. Show that every year, in some month he is paid three times.

16. Suppose that fifty-one small insects are placed inside a square of side 1. Prove that at any moment there are at least three insects which can be covered by a single disk of radius $1/7$.

17. Prove that in any convex $2n$-gon, there is a diagonal not parallel to any side.

18.   (a) Prove that among 11 distinct positive integer numbers, there are two numbers $a < b$ such that the difference $b - a$ ends with 0 (i.e. has the units digit 0).

    (b) Is the above statement true for the tens digit?

19. Every block of a 3×7 board is colored either black or white. Prove that no matter how the board is colored, it contains a rectangle consisting of more than one row and more than one column whose four corners have the same color.

20. Every block of a 5×41 board is colored in one of four colors. Prove that, no matter how the board is colored, there exists at least one same-color-corner rectangle (as in problem 19).

21. Let n be a positive integer which is not divisible by 2 or 5. Prove that there is a multiple of n consisting entirely of ones.

22. One thousand coins of radius 1 cm each are placed on a table of size 55 cm $\times$60 cm. Prove that it is possible to put another coin so that it does not touch any of the 1000 original coins. (The coin must lie completely on the surface of the table, i.e. it cannot stick out.)

23. Prove that if $1 = a_1 < a_2 < a_3 < \ldots < a_8 = 100$ and all a_i's are integers, then $a_{i+1} - a_i \geq 15$ for some i.

24. Let twenty distinct positive integers be all less than 70. Prove that among their pairwise differences there are four equal numbers.

25. Let $\{a_1, a_2, \ldots, a_{n+1}\}$ be numbers from the set $\{1, 2, \ldots, 2n\}$. Prove that at least two of the a_i's are relatively prime.

26. Let $\{a_1, a_2, \ldots, a_{n+1}\}$ be numbers from the set $\{1, 2, \ldots, 2n\}$. Prove that one of the a_i's is divisible by another.

27. Let f be a one-to-one function from $X = \{1, 2, \ldots, n\}$ onto X. Let $f^k = \underbrace{f \circ f \circ \ldots \circ f}_{k \text{ times}}$ denote the k-fold composition of f with itself. Show that for some positive integer m, $f^m(x) = x$ for all $x \in X$.

Chapter 6

Number theory

In this chapter we recall basic properties of integers. (Some of them have been used in previous chapters.)

Note. **All numbers discussed in this chapter are integers.**

Definition 6.1. If $a \neq 0$ and $b = aq$, then we say that a **divides** b, and write $a|b$, or that b is **divisible** by a, and we write $b \vdots a$.

Theorem 6.2. *(Fundamental properties of divisibility)*

1. *If $a|b$ and $b|c$, then $a|c$.*

2. *If $a|b$ and $a|c$, then $a|(bx + cy)$ for any x and y.*
   Important special cases:

   (a) *If $a|b$, then $a|bx$ for any x.*
   (b) *If $a|b$ and $a|c$, then $a|(b + c)$ and $a|(b - c)$.*

Definition 6.3. Let $a \neq 0$ or $b \neq 0$. The largest number that divides both a and b is called **the greatest common divisor** of a and b, and is denoted by $\gcd(a, b)$ or just (a, b).

Theorem 6.4. *For any nonzero numbers a and b, $(a, a) = a$, $(a, 1) = 1$, $(a, 0) = a$, $(a, b) = (b, a)$.*

Definition 6.5. An integer greater than 1 is called **prime** if it has exactly two positive divisors, 1 and itself. An integer greater than 1 that is not prime is called **composite**.

Theorem 6.6. *There are infinitely many primes.*

Theorem 6.7. *(Euclid's lemma) If p is prime, $p|ab$, then $p|a$ or $p|b$.*

Theorem 6.8. *(Fundamental theorem of arithmetic) Every positive integer larger than 1 has a prime factorization, i.e. can be written as a product of primes, and such a product is unique up to order of the factors.*

Example 6.9. $12 = 2 \cdot 2 \cdot 3 = 2 \cdot 3 \cdot 2 = 3 \cdot 2 \cdot 2$ are the only prime factorizations of 12. The order of the factors is different, but the set of factors is the same.

Corollary 6.10. *If $p|a$, $q|a$, and p and q are distinct primes, then $(pq)|a$.*

Definition 6.11. Integers a and b are called **relatively prime**, or coprime, if $(a, b) = 1$.

Remark. The numbers a and b may be relatively prime even if they are both composite. For example, $8 = 2 \cdot 2 \cdot 2$ and $15 = 3 \cdot 5$ are composite but relatively prime since they do not share any factors.

Theorem 6.12. *For any pair of nonzero numbers a and b, their greatest common divisor (a, b) is a linear combination of a and b, i.e. there exist integers x and y such that $(a, b) = ax + by$. Moreover, (a, b) is the smallest positive integer that can be written in the form $ax + by$ for some x and y.*

Corollary 6.13. *Nonzero numbers a and b are relatively prime if and only if there exist x and y such that $ax + by = 1$.*

Definition 6.14. For every pair a, $b \neq 0$ there exist unique q and r such that

$$a = bq + r, \qquad 0 \leq r < b.$$

The numbers q and r are called the **quotient** and **remainder** respectively upon division of a by b.

The following observations are often useful when solving problems involving integers.

- Any integer can be written in the form $10q + r$ for some q and r where $0 \leq r \leq 9$.

- An integer $\underline{a_n a_{n-1} \ldots a_1 a_0}$ (with digits a_n, a_{n-1}, ..., a_1, a_0) can be written as $10^n a_n + 10^{n-1} a_{n-1} + \ldots + 10^2 a_2 + 10 a_1 + a_0$.

If two numbers have the same remainder upon division by b, then they can be written as $bq_1 + r$ and $bq_2 + r$. Their difference is $b(q_1 - q_2)$, and thus it is divisible by b. As was discussed in chapter 5, since there are b possible remainders upon division by b, given $b + 1$ or more numbers, by Dirichlet's Box Principle at least 2 of them have the same remainder. Their difference is divisible by b.

Definition 6.15. Integers a and b are said to be **congruent** mod m, and we write $a \equiv b \pmod{m}$ if $m | (a - b)$. Equivalently, $a - b = mq$ for some q, or $a = b + mq$, or a and b have the same remainder upon division by m.

Example 6.16. $22 \equiv 7 \pmod 5$ because $5 | (22 - 7)$. Note that 22 and 7 have the same remainder upon division by 5.

Theorem 6.17. *(Properties of congruences)*

1. *If $a \equiv b \pmod m$ and $c \equiv d \pmod m$, then $a \pm c \equiv b \pm d \pmod m$ and $ac \equiv bd \pmod m$.*

2. *If $a \equiv b \pmod m$ then $a^c \equiv b^c \pmod m$.*

3. *If $(c, m) = 1$ and $ca = cb \pmod m$, then $a \equiv b \pmod m$.*

Theorem 6.18. *(Fermat's theorem) If p is prime, then $a^p \equiv a \pmod p$.*

Corollary 6.19. *If p is prime and p does not divide a, then $a^{p-1} \equiv 1 \pmod p$ for any a.*

Example 6.20. Let $p = 5$. Then $1^4 = 1 \equiv 1 \pmod 5$, $2^4 = 16 \equiv 1 \pmod 5$, $3^4 = 81 \equiv 1 \pmod 5$, $4^4 = 256 \equiv 1 \pmod 5$, etc.

Finally, we give two useful formulas.

- $a^n - b^n = (a - b)(a^{n-1} + a^{n-2}b + \ldots + ab^{n-2} + b^{n-1})$

- if n is odd, $a^n + b^n = (a+b)(a^{n-1} - a^{n-2}b + \ldots + (-1)^{n-2}ab^{n-2} + (-1)^{n-1}b^{n-1})$

Problems

1. Show that $\sqrt[3]{25}$ is irrational.

2. Show that $\log_2 5$ is irrational.

3. (a) Prove that a natural number is divisible by 9 if and only if the sum of its digits is divisible by 9.

   (b) Prove that if the sum of the digits of a number is 66 then it is not a perfect square (the square of an integer).

4. Show that 3 divides both a and b if and only if 3 divides $a^2 + b^2$.

5. (a) If c is a perfect square, what are the possible values of its units digit?

   (b) Conclude that a number ending with 3 cannot be a perfect square.

6. (a) If c is a perfect square, what are the possible values of its remainder upon division by 4?

   (b) Conclude that a number ending with 66 cannot be a perfect square.

7. Can a number ending with 65 be a perfect square?

8. The four-digit number $\underline{aabb}$ is a perfect square. Find it.

9. Find $2^{100} \bmod 5$ (that is, find the remainder upon division of 2^{100} by 5).

10. The number 8^{2460} is written on a blackboard (it contains over 2,000 digits, so we hope that the reader doesn't mind that we didn't write it out here). The sum of its digits is calculated, then the sum of the digits of the result is calculated and so on, until we get a single digit. What is this digit?

11. Show that $2^{457} + 3^{457}$ is divisible by 5.

12. Show that $A = 3^{105} + 4^{105}$ is divisible by 7. Find $A \bmod 11$ and $A \bmod 13$.

13. Show that if the units digit of a natural number n is 3, then $5 \mid (n^2 + 1)$.

14. Show that for for any integer n, $6 \mid (n^3 + 5n)$.

15. Show that if n is composite, then $2^n - 1$ is composite.

16. Show for any $n \in \mathbb{N}$, 2^n does not divide $n!$.

17. Find all integral solutions of $x + y = xy$.

18. Find all primes p and q such that $p^2 - 2q^2 = 1$.

19. Show that $x^2 - 3y^2 = 17$ has no integral solutions.

20. Find all integral solutions of $x + y = x^2 - xy + y^2$.

21. How many pairs of positive integers are solutions to the equation $2x + 3y = 100$?

22. How many pairs of positive integers are solutions to the equation $5x + 7y = 1234$?

23. Does there exist a positive integer that starts with 123 and is divisible by 4567? If so, find it.

24. Does there exist a positive integer that ends with 123 and is divisible by 4567? If so, find it.

25. Do there exist integer numbers m and n such that $m^2 + 12345678 = n^2$?

26. Prove that every odd integer can be written as the difference of two perfect squares.

27. A couple has a daughter and a son. The husband is 3 years older than his wife, and their daughter is 2 years older than their son. The sum of ages of all four members of this family is 73 years. Four years ago the sum of all members of the family was 58 years. What are the ages of all four people now?

28. A store manager made 24 bags of apples, some 5 kgs and others 3 kgs. The combined weight of 5 kgs bags was equal to the combined weight of 3 kgs bags. How many bags of each size did he make?

29. Suppose we write a natural number at each vertex of a cube. Then at the midpoint of each edge we write the sum of the two numbers that are at the ends of this edge. Finally, in the middle of each face we write the sum of the four numbers that are at the vertices of this face. Could the sum of all 26 numbers be equal to 1234?

30. Prove that a natural number that consists of 300 ones, some zeros, and no other digits, cannot be a perfect square.

31. The product of four consecutive integers is 3024. Find these integers.

32. Prove that the sum of four consecutive odd integers is divisible by 8.

33. When 3 was appended to a three-digit number on the left (i.e. written before the number), the number increased by 9 times. What was the number?

34. When 6 was appended to a number on the right (i.e. as the rightmost digit), it increased by 13 times. What was the number?

35. When 36 was appended to a number on the right, it increased by 103 times. What was the number?

Chapter 7

Case analysis

We have seen in the previous chapter that some number theory problems can be solved by considering all possible remainders mod n for a variable. The problem about roads in Sikinia (problem 7 in chapter 4) also required considering some cases separately. As we will see in future chapters, the technique of considering all possible cases can be used in many different problems of very different types.

Below are given two typical situations where we need to consider two or more cases.

If $a, b \in \mathbb{Z}$, then $a^b = 1$ if and only if at least one of the following holds:

- $a = 1$,

- $a \neq 0, b = 0$,

- $a = -1$ and b is even.

Example 7.1. Find all integer values of x for which $(x^2 - 5x + 5)^{x^2 - 9x + 20} = 1$.

Solution. *Let $a = x^2 - 5x + 5$ and $b = x^2 - 9x + 20$. Then we have $a^b = 1$.*

Case I: $a = 1$, then $x^2 - 5x + 5 = 1$. Therefore $x = 1$ or $x = 4$.

Case II: $a \neq 0$, $b = 0$. First solve $b = 0$, i.e. $x^2 - 9x + 20 = 0$. Then $x = 4$ or $x = 5$. For both roots $a \neq 0$.

Case III: $a = -1$ and b is an even integer. First solve $a = -1$, i.e. $x^2 - 5x + 5 = -1$. Then $x = 2$ or $x = 3$. For both roots b is even.

Thus solutions are 1, 2, 3, 4, and 5.

Recall that the absolute value of a real number x is denoted $|x|$ and is given by

$$|x| = \begin{cases} x & \text{if} \quad x \geq 0 \\ -x & \text{if} \quad x < 0 \end{cases}$$

Here is the graph of $|x|$:

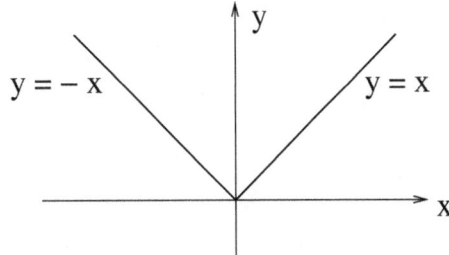

To solve problems involving the absolute value, consider the following two cases: when the expression inside the absolute value is positive or 0, and when it is negative.

Example 7.2. Solve $3|x^2 - 9| - 11x + 7 = 0$.

Solution. *Case I: if $x^2 - 9 \geq 0$, then $|x^2 - 9| = x^2 - 9$, and the equation becomes $3(x^2 - 9) - 11x + 7 = 0$, or $3x^2 - 11x - 20 = 0$. Using the quadratic formula, we find $x = 5$ or $x = -4/3$. The root $x = 5$ satisfies the condition $x^2 - 9 \geq 0$, but $x = -4/3$ does not, so we throw it away.*
Case II: if $x^2 - 9 < 0$, then $|x^2 - 9| = -(x^2 - 9)$, and the equation becomes, $-3(x^2 - 9) - 11x + 7 = 0$, or $-3x^2 - 11x + 34 = 0$. The roots are 2 and $-17/3$, but the second root does not satisfy the condition $x^2 - 9 < 0$, so we throw it away.
Thus the only roots are 5 and 2.

If there are several absolute values, we can consider two cases for each absolute value. Note that if there are n absolute values, then we have 2^n cases total. An alternative, often shorter, way is to find all the points at which the expressions inside absolute values change sign. These points divide the real number line into several intervals (two of which are infinite). Consider each interval separately.

Problems

1. Prove that if n is an integer, then $n^2 + 2$ is not divisible by 5.

2. Solve for x over $\mathbb{Z}$:

   (a) $x^{x^2 - 7x + 12} = 1$

   (b) $(x - 3)^{x^2 - 8x + 15} = 1$

   (c) $(x^{x+1})^{x^2} = 1$

3. Solve for x over $\mathbb{Z}$:

   (a) $x^{(x^2)} = x^2$

   (b) $x^{((x+1)^2)} = x^{16}$

   (c) $x^{(x^x)} = (x^x)^x$

   (d) $\sqrt{x^{x+1}} = x^{\sqrt{x+1}}$

4. Find all pairs (x, y) of integers that satisfy the system $\begin{cases} x^{2x} &= y + 1 \\ x^y &= 1. \end{cases}$

5. Find all pairs (x, y) of integers that satisfy the system $\begin{cases} x^{x+y} &= y^4 \\ y^{x+y} &= x. \end{cases}$

6. Solve for x over $\mathbb{R}$: $x^2 + |2x - 2| = 1$

7. Solve for x over $\mathbb{R}$:

   (a) $|2x + 3| - |x| = 3$

   (b) $|2x - 1| - |x + 5| = 3$

   (c) $|3x + 6| + |x - 1| = 2$

   (d) $|x + 1| + |3x + 1| = |2x + 1| + |4x + 1|$

8. Solve for x over $\mathbb{R}$:

   (a) $x^2 - |5x - 6| \leq 0$

   (b) $|x + 1| + 5 - x^2 > 0$

   (c) $x^2 - |7x + 15| \geq 3$

9. Solve for x over $\mathbb{R}$:

   (a) $|x - 5| + |2x - 4| \leq 6$

   (b) $|x - 5| - |x - 2| \geq 1$

   (c) $|x - 1| - |x - 3| > 5$

10. Sketch the graph of

    (a) $f(x) = |x^2 - 4| + 2$

    (b) $g(x) = |x^2 - 1| - |x^2 - 4|$

11. Sketch the graph of

    (a) $f(x) = |x + |x + 2||$

    (b) $g(x) = |x^2 - 4|x| + 3|$

12. Sketch the graph of $|x| + |y| = 1 + |xy|$.

13. Sketch the region:

    (a) $\{(x, y) \mid |x| + |y^3| < 8\}$

    (b) $\{(x, y) \mid |x - y| + |x| - |y| \leq 2\}$

    (c) $\{(x, y) \mid 2|y - x| + |y + x| \leq 1\}$

14. Find all integral solutions of

    (a) $a^b = 625$

    (b) $(a^b)^c = 64$

Chapter 8

Finding a pattern

As suggested in the hint for problem 14 of chapter 4, a formula can be guessed after computing a few values. Although a guess is not sufficient and a rigorous proof is needed to ensure correctness of a formula, guessing often is a powerful technique. In this chapter we will consider problems in which we can find the first few values of a certain sequence, guess a formula for a general case, and then prove it (e.g. by Mathematical Induction or some other proof tool).

Example 8.1. Find the n-th derivative of $f(x) = 5^x$.

Solution. *Find the first few derivatives (until you can see a pattern):*
$f'(x) = \ln 5 \cdot 5^x$
$f''(x) = \ln 5 \cdot \ln 5 \cdot 5^x = (\ln 5)^2 \cdot 5^x$
$f'''(x) = (\ln 5)^2 \cdot \ln 5 \cdot 5^x = (\ln 5)^3 \cdot 5^x$
We notice that $f^{(n)}(x) = (\ln 5)^n \cdot 5^x$.
It is easy to prove this formula using Mathematical Induction.
The basis step is $f'(x) = \ln 5 \cdot 5^x$.
Inductive step: suppose $f^{(k)}(x) = (\ln 5)^k \cdot 5^x$ *is true. Then*
$f^{(k+1)}(x) = (f^{(k)}(x))' = ((\ln 5)^k \cdot 5^x)' = (\ln 5)^k \cdot \ln 5 \cdot 5^x = (\ln 5)^{k+1} \cdot 5^x$.

Example 8.2. Guess a formula for the n-th term of the sequence: 1, 3, 6, 10, 15, 21, ...

Solution. *Notice that the difference between the first and the second terms is 2, the difference between the second and the third terms is 3, and then the differences are 4, 5, 6, ... Thus*
$a_1 = 1$,
$a_2 = 1 + 2$,
$a_3 = 1 + 2 + 3$,
$a_4 = 1 + 2 + 3 + 4$,
$a_5 = 1 + 2 + 3 + 4 + 5$,
$a_6 = 1 + 2 + 3 + 4 + 5 + 6$.
So it appears that $a_n = 1 + 2 + \ldots + n = \dfrac{n(n+1)}{2}$.

Note. For the problems like this one, since only a few terms of a sequence are given, there may be several different formulas valid for these few terms. For example, $a_n = 2^n - \dfrac{1}{60}n^5 + \dfrac{1}{6}n^4 - \dfrac{11}{12}n^3 + \dfrac{7}{3}n^2 - \dfrac{77}{30}n$ is another valid formula for this sequence. But we tried to find a simple one.

Problems

1. Guess a formula for the n-th term of the sequence $a_1, a_2, a_3, \ldots$ whose first few terms are given. Try to find the simplest possible formula, but any correct formula (that is, any formula that works for the given terms) is acceptable.

    (a) 1, 4, 9, 16, 25, 36, 49, ...

    (b) -1, 0, 1, 2, 3, 4, 5, ...

    (c) 5, 7, 9, 11, 13, 15, ...

    (d) 8, 10, 12, 14, 16, 18, ...

    (e) 3, 1, -1, -3, -5, -7, ...

    (f) 1, 2, 1, 4, 1, 6, 1, 8, ...

    (g) 1, 3, 4, 6, 7, 9, 10, 12, ...

    (h) 0, 1, 3, 7, 15, 31, ...

    (i) $\dfrac{1}{2}, \dfrac{1}{2}, \dfrac{3}{8}, \dfrac{1}{4}, \dfrac{5}{32}, \dfrac{3}{32}, \dfrac{7}{128}, \ldots$

2. Compute $S_n = \dfrac{1}{1 \cdot 2} + \dfrac{1}{2 \cdot 3} + \dfrac{1}{3 \cdot 4} + \ldots + \dfrac{1}{(n-1)n}$ for some small values of n. Notice the pattern. Write a formula for S_n and prove it using Mathematical Induction.

3. Find a formula for $\dfrac{1}{1 \cdot 3} + \dfrac{1}{3 \cdot 5} + \ldots + \dfrac{1}{(2n-1)(2n+1)}$.

4. Find a formula for

$$\prod_{i=1}^{2n-1}\left(1 - \frac{(-1)^i}{i}\right) = \left(1 - \frac{-1}{1}\right)\left(1 - \frac{1}{2}\right)\left(1 - \frac{-1}{3}\right)\cdots\left(1 - \frac{-1}{2n-1}\right).$$

5. Let $f_1(x) = 2x + 1$ and $f_n = f_1 \circ f_{n-1}$ for $n \geq 2$. Compute f_n for some small values of n. Notice the pattern. Write a closed formula for f_n (i.e. a formula that does not involve f_k for $k < n$). and prove it using Mathematical Induction.

6. Let $f_1(x) = \dfrac{1}{2-x}$ and $f_n = f_1 \circ f_{n-1}$ for $n \geq 2$. Find and prove a closed formula for $f_n(x)$.

7. The units digit of a number a^b can be found by computing the units digits of the first few powers of a, i.e. a^1, a^2, a^3, etc. and noticing a pattern. Find the units digit of the following numbers.

    (a) 107^{107}

    (b) 1234^{5678}

8. Find the last two digits of 7^{50}.

9. Find the remainder of 2^{100} upon division by 12.

10. Find the remainder of 5^{4321} upon division by 11.

11. Find the n-th derivative of

(a) $f(x) = \sin(x)$,

(b) $g(x) = \ln(x)$,

(c) $h(x) = 2e^{5x}$.

12. Suppose that n lines in general position are given in a plane. (General position means that no two lines are parallel, and no three lines have a common point.) Into how many regions do they divide the plane?

13. Suppose that n circles are given in a plane, such that every pair of circles has 2 intersection points, but no 3 circles have a common point. Into how many regions do they divide the plane?

14. Amanda is training her rabbit to climb a flight of 10 steps. The rabbit can hop up 1 or 2 steps each time he hops. He never hops down, only up. In how many different ways can he hop up the flight of 10 steps?

15. Let $F_0 = 0$, $F_1 = 1$, $F_2 = 1$, ..., F_{99} be the first 100 Fibonacci numbers (recall that $F_n = F_{n-1} + F_{n-2}$ for $n \geq 2$).

    (a) How many of them are even?

    (b) How many of them are divisible by 3?

    (c) How many of them are divisible by 5?

Chapter 9

Invariants

Definition 9.1. An **invariant** is something that does not change.

Example 9.2. The numbers $1, 2, \ldots, 10$ are written on the blackboard. We pick any two numbers, let us call them a and b. We erase these numbers, and write $a+1$ and $b-1$ instead. Is it possible to get ten 5's by a sequence of such operations?

Solution. *Notice that when we increase a by 1 and decrease b by 1, the sum of the numbers does not change (so in this example the sum of the ten numbers is an invariant). Initially the sum is $1 + 2 + \ldots + 10 = 55$, and $10 \cdot 5 = 50$, therefore it is not possible to get ten 5's.*

Example 9.3. Each of the numbers $a_1, a_2, \ldots, a_n$ is either 1 or -1, and

$$a_1a_2a_3a_4 + a_2a_3a_4a_5 + \ldots + a_{n-3}a_{n-2}a_{n-1}a_n + a_{n-2}a_{n-1}a_na_1 + a_{n-1}a_na_1a_2 + a_na_1a_2a_3 = 0.$$

Prove that $4|n$.

Solution. *Let*

$$S = a_1a_2a_3a_4 + a_2a_3a_4a_5 + \ldots + a_{n-3}a_{n-2}a_{n-1}a_n + a_{n-2}a_{n-1}a_na_1 + a_{n-1}a_na_1a_2 + a_na_1a_2a_3.$$

If we replace a_i by $-a_i$, then S does not change modulo 4 since four terms (containing a_i) change their sign. Indeed, if all four terms are of the same sign, then their sum changes either from -4 to 4 or from 4 to -4, thus S changes by ± 8. If one or three of these four terms are positive, then the sum of the four terms changes either from -2 to 2 or from 2 to -2, thus S changes by ± 4. Finally, if two of these four terms are positive and two are negative, then the sum does not change. Initially, we have $S = 0$ which implies $S \equiv 0 \pmod 4$. Now, step-by-step, we can change each -1 into a 1. At the end, we have $S = n$, and we must still have $S \equiv 0 \pmod 4$, so $4|n$.

Here are a few things that are very often invariants in problems involving sets of numbers and allowed operations, so you may want to try look at them. Sometimes, of course, you have to be very creative!

- The sum or the product of all given numbers

- The number of positive or negative numbers

- The number of even or odd numbers, or, more generally, the number of numbers congruent to a modulo b for some integers a and b

- One of the above modulo a positive number (e.g. the sum modulo 2, i.e. the parity of the sum; the product modulo 3; the number of positive numbers modulo 4; etc.)

Sometimes a quantity that may change is useful as well. Especially if it can change only in a certain way, say, it can only increase or decrease. For instance, in the example below we find a positive decreasing function rather than a constant function. The idea is that the value of that function must be non-negative. We apply a series of steps each of which decreases the value of the function. Since the value cannot become negative, sooner or later it will reach 0.

Example 9.4. Suppose that $2n$ ambassadors are invited to a banquet. Every ambassador has at most $n - 1$ enemies. Prove that the ambassadors can be seated around a round table so that nobody sits next to their enemy.

Solution. *First, we seat the ambassadors randomly. Let H be the number of neighboring hostile couples. We must find an algorithm which reduces this number whenever $H > 0$. Let (A, B) be a hostile couple with B sitting to the right of A:*

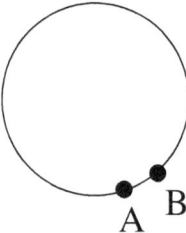

We want to separate them so as to not gain any new neighboring hostile couples. This can be achieved by reversing some arc BC as shown below. The number H will be reduced if (A, C) and (B, D) are friendly couples.

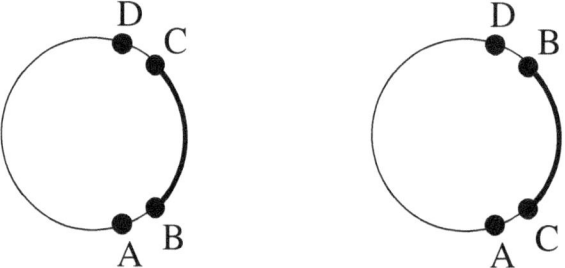

It remains to be shown that such a couple C, D always exists. We start at A and go around the table counterclockwise. We will encounter at least n friends of A. To their right, there are at least n seats. They cannot all be occupied by enemies of B since B has at most $n - 1$ enemies. Thus there is a friend C of A with the right neighbor D being a friend of B.

Problems

1. We start with the set $\{-3, -2, -1, 1, 2, 3\}$. In each step we may choose any two of these numbers and change their signs. We may repeat this step as many times as we want. Show that it is not possible to reach the set $\{3, 2, 1, 1, 2, 3\}$.

2. We start with the set $\{-3, -2, -1, 1, 2, 3\}$. In each step we may multiply or divide any of these numbers by any positive number. We may repeat this step as many times as we want. Show that it is not possible to reach the set $\{-2, -1, 1, 2, 3, 4\}$.

3. We start with the set $\{1, 1, 1, 1\}$. In each step we may either multiply one of the numbers by 3, or subtract 2 from it. We may repeat this step as many times as we want. Is it possible to reach the set $\{1, 2, 3, 4\}$?

4. We start with the set $\{1, 4, 32, 128, 256\}$. In each step we may divide one number by 2 and multiply another number by 2. We may repeat this step as many times as we want. Is it possible to reach the set $\{512, 32, 16, 16, 2\}$?

5. Initially 1 is written in every cell of a 5×5 table. In each step we may change the signs of the numbers in any two adjacent cells. We may repeat this step as many times as we want. Is it possible to make all of the numbers -1?

6. We start with the set $\{1, 2, 3, 4, 5, 6\}$. In each step we may either add 2 to any 5 numbers or subtract 1 from any 5 numbers. We may repeat this step as many times as we want. Can we reach $\{1, 2, 4, 8, 16, 32\}$?

7. We start with the set $\{1, 2, 3, 4\}$. In each step we may either add 2 times one of the numbers to another number or subtract 2 times one number from another number. For example, we may replace 1 by $1 + 2 \cdot 2$, or by $1 - 2 \cdot 2$, or by $1 + 2 \cdot 3$, etc. We may repeat this step as many times as we want. Can we reach $\{10, 20, 30, 40\}$?

8. We start with the table shown below. In one step, we may either add 1 to all the numbers in any row or column, or subtract 1 from all the numbers in any row or column. We may repeat this step as many times as we want. Prove that it is not possible to reach nine 1's.

0	0	0
0	1	0
0	0	0

9. There are several $+$ and $-$ signs written on a board. We may erase any two signs and write, instead, $+$ if they are equal and $-$ if they are unequal. We do this until only one sign remains. Prove that the last sign on the board does not depend on the order of erasure.

10. Assume we have an 8×8 chessboard with the usual coloring. We may switch the colors of all the squares in any row or column. We may repeat this step as many times as we want. The goal is to attain just one black square. Can we reach the goal? What if we are allowed to switch the colors of all the squares in any 2×2 square?

11. Each of the numbers from 1 to 10^6 is repeatedly replaced by its digital sum until we reach 10^6 one-digit numbers. For example, 987654 is replaced by 39 (since $9 + 8 + 7 + 6 + 5 + 4 = 39$), then 39 is replaced by 12 (since $3 + 9 = 12$), and finally, 12 is replaced by 3 (as $1 + 2 = 3$). Among these 10^6 one-digit numbers, will we have more 1's or 2's?

12. We may write all the digits from 1 to 9 in a row in any order we like, and then we write plus signs between some digits (as many plus signs as we like). For example, we could write $7 + 35 + 19 + 4 + 286$. Finally, we evaluate the obtained expression. Prove that there is no way to get the value of 100. Or 101. Or 102. Or 103... What is the smallest three-digit number that can be obtained in this game?

13. Let n be any positive integer. We start with the integers $1, 2, \ldots, 4n - 1$. In each step we may replace any two integers by their difference. We do this until only one number remains. Prove that an even integer will be left after $4n - 2$ steps.

14. Let n be an odd positive integer. We start with the numbers $1, 2, 3, \ldots, 2n$. In each step we may replace any two integers by their difference. We do this until only one number remains. Prove that an odd number will remain at the end.

15. The numbers from 0 to 9 are written along a circle in random order. Between every two neighboring numbers a and b (in the clockwise order) we write $2b - a$. Then we erase the original ten numbers. (For example, the numbers could be written in the following order: 1, 5, 3, 9, 0, 2, 4, 6, 8, 7. Then the new numbers would be 9, 1, 15, -9, 4, 6, 8, 10, 6, -5.) This step can be repeated as many times as we want. Show that it is not possible to reach ten 5's.

16. We start with the set $\{1, 3, 6\}$. In each step we may choose any two of these numbers, let's call them a and b, and replace them by $0.6a - 0.8b$ and $0.8a + 0.6b$. We may repeat this step as many times as we want. Can we reach the set $\{2, 4, 5\}$?

17. We arrange the integers $1, 2, 3, 4, 5, 6$ in any order on six places numbered 1 through 6. Next we add each number to its place. Prove that at least two of the sums have the same remainder upon division by 6.

18. There are seven 1's and eight -1's on a board. In each step we may erase any two numbers, say, a and b, and write $-ab$ instead. We do this until only one number remains. Show that no matter in what order we erase the numbers, 1 will remain in the end.

19. Nine 1×1 cells of a 10×10 square are infected. Two cells are called neighbors if they have a common side. In one time unit, the cells with at least two infected neighbors become infected. Can the infection spread to the whole square (in any amount of time)?

20. Twelve 1×1 cells of a 10×10 square are infected. Two cells are called neighbors if they share at least one vertex (thus an inner cell has 8 neighbors). In one time unit, the cells with at least four infected neighbors become infected. Can the infection spread to the whole square (in any amount of time)?

21. In the Congress of Sikinia each member has at most three enemies. Prove that the congress can be separated into two houses so that each member has at most one enemy in his own house.

22. There are a white, b black, and c red chips on a table. In one step we may choose two chips of different colors and replace them by one chip of the third color. We do this until all remaining chips are of the same color. If only one chip remains, prove that its color does not depend on the evolution of the game, but it only depends on the numbers a, b, and c.

23. A circle is divided into six sectors. Then the numbers 1, 0, 1, 0, 0, 0 are written into the sectors as shown below. We may increase any two neighboring numbers by 1. We may repeat this step as many times as we want. Is it possible to equalize all the numbers?

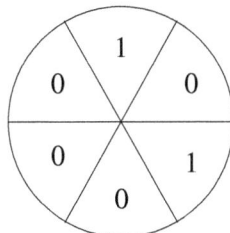

24. In the table below, we may switch the signs of all the numbers in a row, column, or parallel to one of the diagonals. In particular, we may switch the sign of each corner square. We may repeat this step as many times as we want. Prove that at least one -1 will remain in the table.

-1	1	-1	1
1	1	1	1
1	1	-1	-1
1	-1	1	1

Chapter 10

Coloring

In all examples and problems in this chapter, "cover" means cover without overlap.

Example 10.1. In 1961, the British theoretical physicist M. E. Fisher solved a famous and very tough problem. He showed that an 8×8 chessboard can be covered by 2×1 dominoes in $2^4 \cdot 901^2 = 12{,}988{,}816$ ways. Now let us cut out two diagonally opposite corners of the board. In how many ways can we cover the 62 squares of the mutilated chessboard with 31 dominoes?

Solution. *Zero. There is no way to cover the mutilated chessboard. Each domino covers one black and one white square. If a covering of the board existed, it would cover 31 black and 31 white squares. But the mutilated chessboard has 30 squares of one color and 32 squares of the other color.*

Example 10.2. A rectangular floor is covered by 2×2 and 4×1 tiles. One tile got smashed. There is a tile of the other kind available. Show that the floor cannot be covered by rearranging the tiles.

Solution. *Let us color the floor as shown in the picture below. A 4×1 tile always covers either 0 or 2 black squares. A 2×2 tile always covers one black square. Therefore it is impossible to exchange one tile for a tile of the other kind.*

Besides the colorings used in the above examples, the "stripe colorings" and the "diagonal colorings" shown below are often helpful.

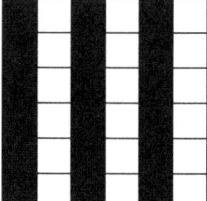 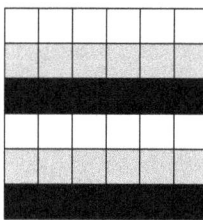  

Also, you can use the stripe or diagonal pattern with more colors; or the stripe pattern with one stripe of one color followed by several stripes of another color; and so on. Of course, sometimes you have to be creative and find your own coloring that would work for a particular problem!

Problems

1. Prove that a 14 × 14 board cannot be covered by 49 T-tetrominoes (see pictures of tetrominoes below).

2. Prove that an 8 × 8 board cannot be covered by 15 T-tetrominoes and one square tetromino.

3. Prove that a 10×10 board cannot be covered by 15 T-tetrominoes and 10 L-tetrominoes.

4. Is it possible to form a rectangle with the five tetrominoes shown below (using one tetromino of each kind)?

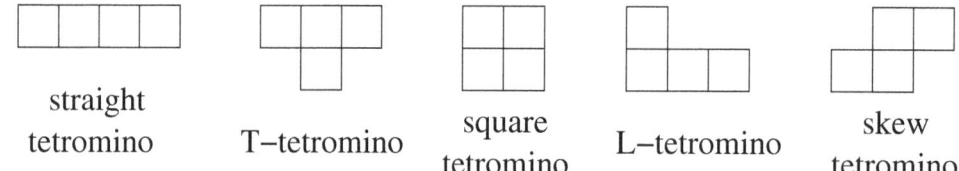

straight
tetromino   T−tetromino   square   L−tetromino   skew
                                        tetromino   tetromino

5. An 8 × 8 board is covered by tetrominoes. Prove that the number of T-tetrominoes is even.

6. Is it possible for a chess knight to start at the upper left corner, go through every square on the chessboard exactly once and reach the lower right corner? (See allowed moves in chapter 1.)

7. Prove that the figure shown below (with the center block removed) cannot be covered by dominoes.

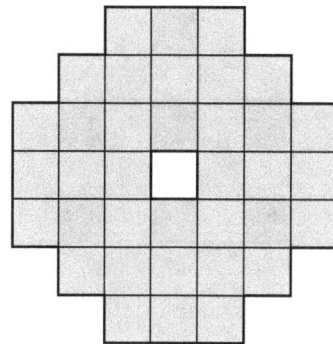

8. The figure below shows a road map connecting 14 cities. Is there a path passing through each city exactly once?

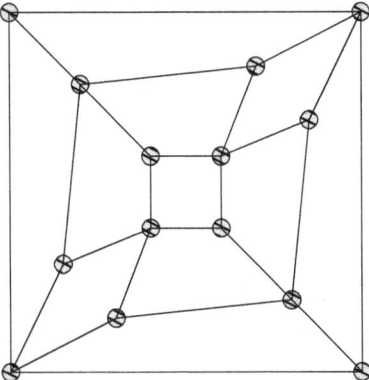

9. Prove that a 6×6 board cannot be covered by 9 L-tetrominoes.

10. Prove that an 8×8 board cannot be covered by 11 straight tetrominoes and 5 L-tetrominoes.

11. Prove that an 8×8 board with one corner square removed (so, 63 squares remain) cannot be covered by 21 straight trominoes (i.e. 3×1 tiles).

12. Prove that a 15×8 board cannot be covered by 2 L-tetrominoes and 28 skew tetrominoes.

13. Prove that a 23×23 board cannot be covered by 2×2 and 3×3 tiles.

14. Prove that a 10×10 board cannot be covered by 25 straight tetrominoes.

15. Prove that an $a \times b$ rectangle can be covered by $1 \times n$ rectangles if and only if $n|a$ or $n|b$.

16. A 7×7 board is covered by sixteen 3×1 and one 1×1 tiles. What are the possible positions of the 1×1 tile?

17. (a) The vertices and midpoints of the faces are marked on a cube, and all face diagonals are drawn. Prove that there is no path along the face diagonals that visits each marked point exactly once.

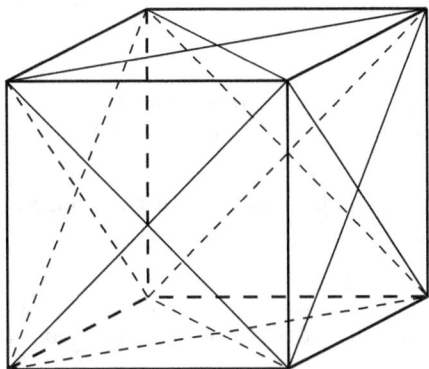

(b) Show that if one walk along an edge is allowed, then there is a path visiting all the marked points. (Find such a path.)

18. The map below shows the cities and one-way roads in Sikinia.

(a) Prove that there is no closed path (a path is closed if it starts and quits in the same city) that visits every city exactly once.

(b) Is there a closed path that visits every city exactly twice?

(c) Is there a path, not necessary closed, that starts in the upper left corner and visits every city exactly once?

(d) Is there a path, not necessarily closed, that starts in the upper left corner and visits every city exactly twice?

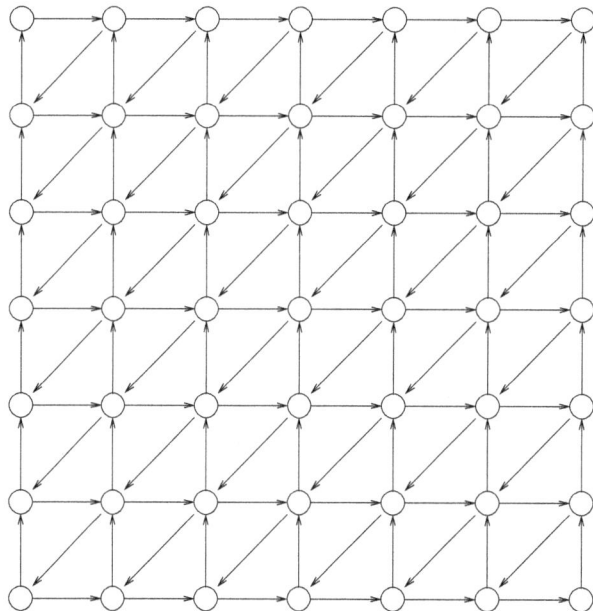

19. Prove that there is no way to pack fifty-four $1 \times 1 \times 4$ bricks into a $6 \times 6 \times 6$ box.

20. Show that if $4 \times 1 \times 1$ bricks and $2 \times 2 \times 2$ cubes fill an $8 \times 8 \times 8$ cube, then the number of $2 \times 2 \times 2$ cubes is even.

21. Is it possible for a chess knight to pass through all the squares of a 123×4 board having visited each square exactly once, and return to the initial square?

22. Is it always possible to cover a chessboard with two squares removed, one black and one white, by 31 dominoes?

23. Is it possible to write distinct natural numbers from 1 to 16 in the small triangles in the figure below so that the sum of the two numbers in any two triangles that have a common side is prime?

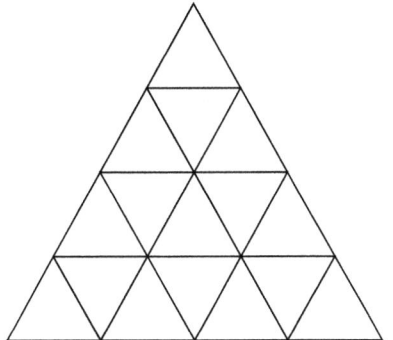

Chapter 11

Areas and Volumes

Recall the following area and volume formulas:

1. Triangle

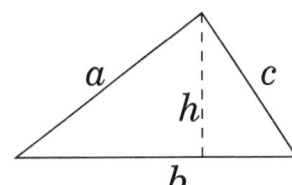

Area $A = \frac{1}{2}bh = \sqrt{\frac{p}{2}\left(\frac{p}{2} - a\right)\left(\frac{p}{2} - b\right)\left(\frac{p}{2} - c\right)}$
where $p = a + b + c$.

2. Trapezoid

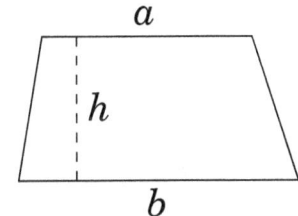

Area $A = \frac{1}{2}(a + b)h$.

3. Ball

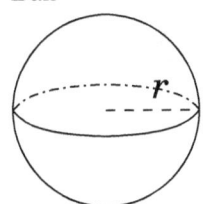

Volume $V = \frac{4}{3}\pi r^3$
where r is the radius.

4. Pyramid and cone

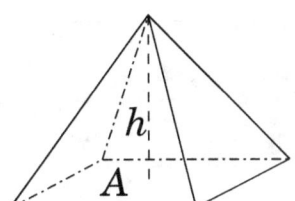 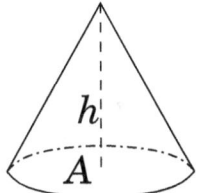

Volume $V = \frac{1}{3}Ah$
where A is the area of the base.

Problems

For problems 1-16, find the area of the given (shaded) region. If a grid is given, assume that each small square is 1×1.

1.

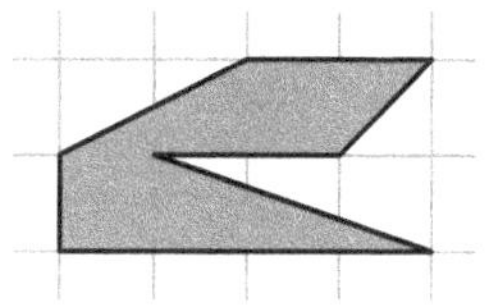

2 .

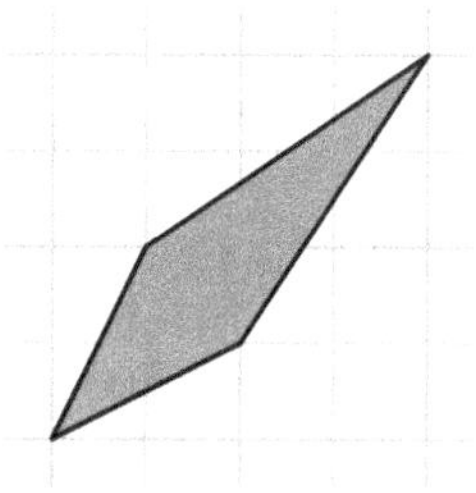

3.

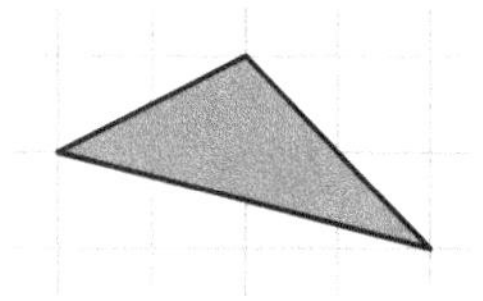

4. Regular octagon

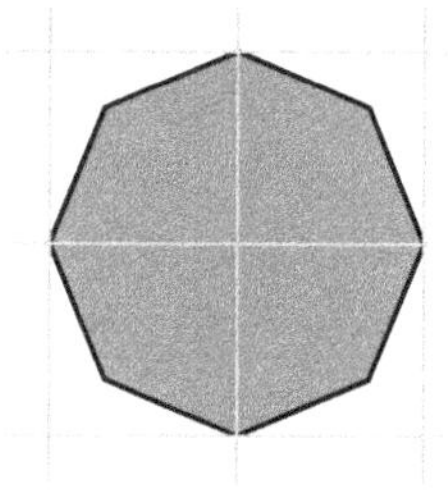

5.

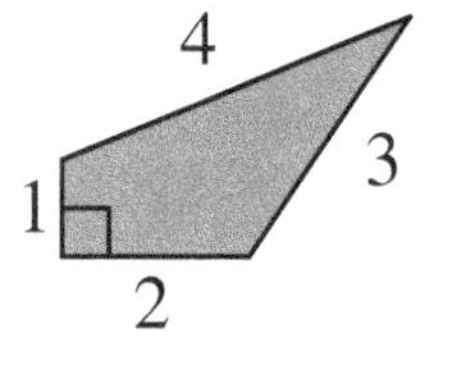

6.

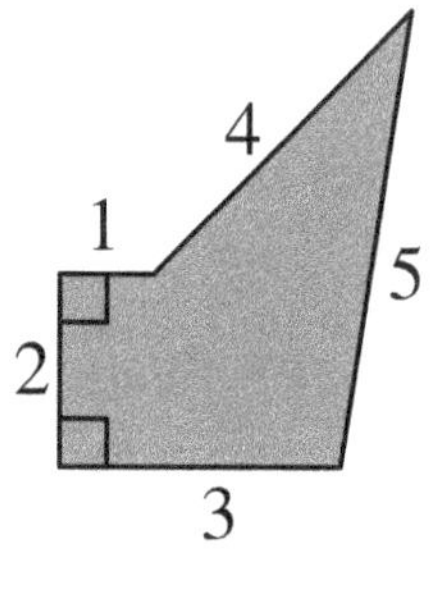

7. Trapezoid

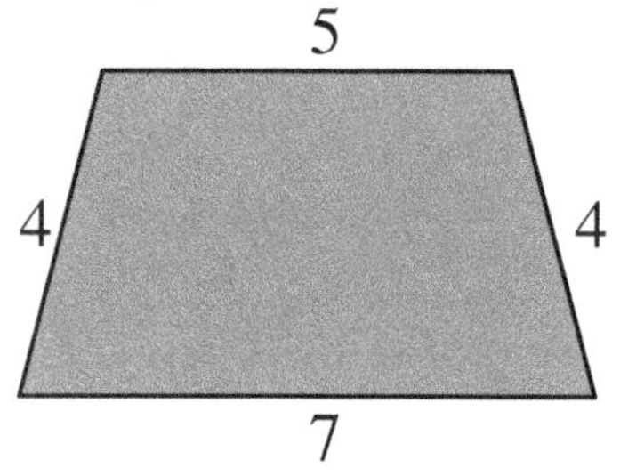

8. Trapezoid

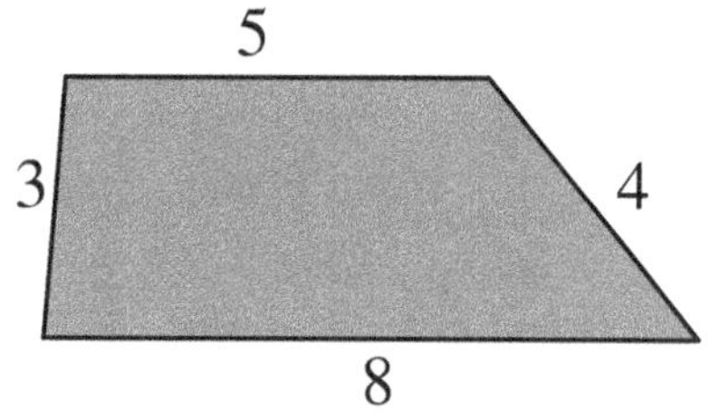

In problems 9 and 10 the curves that appear to be arcs of circles are indeed arcs of circles.

9.

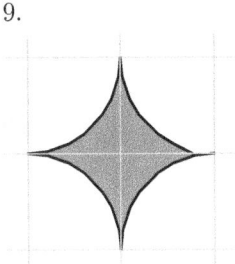

10.

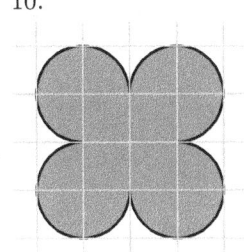

In problems 11-14 each circle has radius 1 and passes through the center of each other circle.

11.

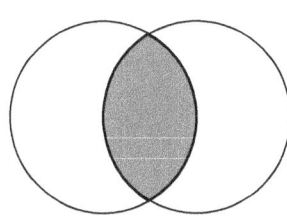

12.

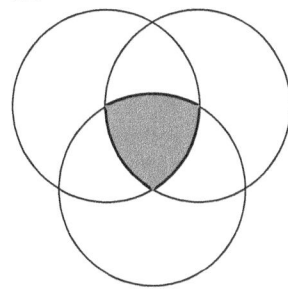

13.

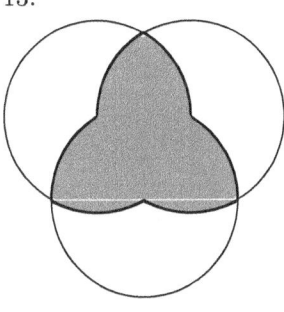

14.

15.

16. Lines are tangent to the circle

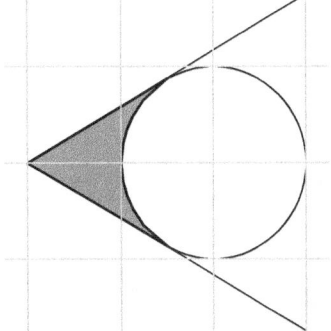

17. Sketch the region $S = \{(x, y) \mid |x| + 2|y| \leq 4\}$ and find its area.

18. Sketch the region $S = \{(x, y) \mid x \geq 2, x^2 + y^2 \leq 16\}$ and find its area.

19. An open box is formed from a square of cardboard by cutting a 3 cm × 3 cm square from each corner and folding up the edges to form the sides. If the volume of the box is 60 cubic cm, find the dimensions of the box.

20. What are the dimensions of a box with square bottom and open top if its volume is 500 cubic cm and its surface area of 300 square cm?

21. Find the volume of a regular octahedron with edge of length 1.

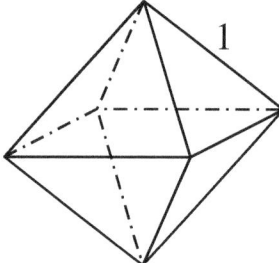

22. Find the volume of a truncated pyramid with a 2 × 2 base, a 1 × 1 top, and side edge of length 1).

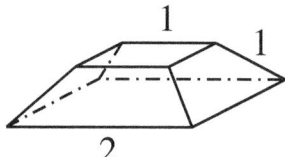

23. Find the volume of the solid inside a sphere of radius 1 and above the cube inscribed in it.

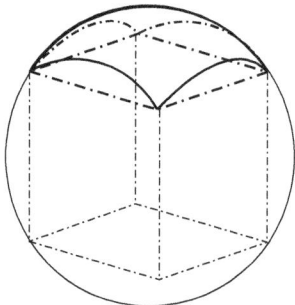

Chapter 12

Symmetry, Translations, Rotations, and Similarity

The following geometry problem can be solved algebraically as well as geometrically.

Example 12.1. There are two poles of heights a and b as shown below. The distance between the poles is d. Find the point on the ground equidistant from the tops of the poles.

Depending on our goal, the word "find" here could mean either "calculate the location of that point" (e.g. "find the distance between the point and one of the poles"), or "construct this point geometrically". Say, suppose we have a geometry software such as *Geometer's sketchpad* that can do the following things: translate objects any given distance in any given direction; rotate objects about any given point through any given angle; reflect objects about any given line; draw parallel and perpendicular lines; and bisect line segments and angles. Note that all these operations can be done on paper using a ruler and a compass.

Now we will solve the above problem.

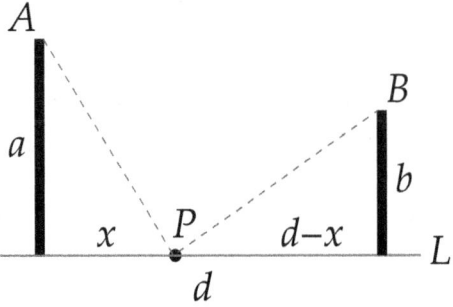

Calculation. *Let x be the distance between one of the poles and the point P we are looking for. Then the distance between the other pole and the point P is $d-x$. We use the Pythagorean theorem to compute the distances between P and the tops of the poles A and B, and set the two distances equal: $\sqrt{a^2 + x^2} = \sqrt{b^2 + (d-x)^2}$. Squaring both sides gives $a^2 + x^2 = b^2 + (d-x)^2$, or $a^2 = b^2 + d^2 - 2dx$. Therefore $2dx = b^2 + d^2 - a^2$, so $x = \dfrac{b^2 + d^2 - a^2}{2d}$.*

Construction. *The point P is equidistant from the tops of the poles iff it lies on the perpendicular bisector of AB. Thus all we have to do is to draw the perpendicular bisector of AB, and then P is its intersection with the line L.*

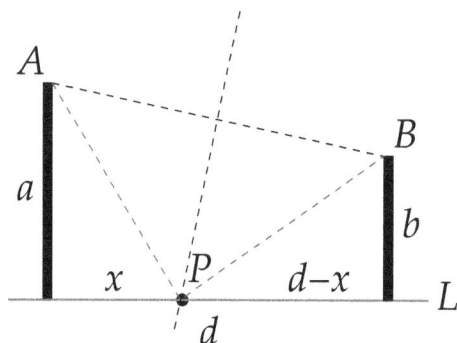

Remark. If we do want to calculate the position of the point P, it is easy to do so using the above construction. Let us introduce a coordinate system with the origin, say, at the bottom of the left pole; find the coordinates of the midpoint of AB; find the slope of the line AB; write an equation of the perpendicular bisector, and find its x-intercept. Namely, the coordinates of the midpoint of AB are $\left(\dfrac{d}{2}, \dfrac{a+b}{2}\right)$. The slope of the line AB is $\dfrac{b-a}{d}$, and therefore the slope of a perpendicular line is $-\dfrac{d}{b-a}$. Therefore an equation of the perpendicular bisector is $y - \dfrac{a+b}{2} = -\dfrac{d}{b-a}\left(x - \dfrac{d}{2}\right)$. Its x-intercept has $y = 0$, therefore $-\dfrac{a+b}{2} = -\dfrac{d}{b-a}\left(x - \dfrac{d}{2}\right)$. This implies $x - \dfrac{d}{2} = \dfrac{(a+b)(b-a)}{2d}$, so $x = \dfrac{(a+b)(b-a)}{2d} + \dfrac{d}{2} = \dfrac{b^2 - a^2 + d^2}{2d}$.

However, if our goal is to find the point geometrically, it is much easier to draw the perpendicular bisector of AB than solve an equation and construct a line segment of length $\dfrac{b^2 + d^2 - a^2}{2d}$ (although the latter is also possible). In some problems even finding all required distances is easier by a calculation based on a geometric construction than purely algebraically. Consider the following example.

Example 12.2. A rope of length l is strung between the tops of two poles of heights a and b, and a weight is hung from a ring on the rope. The rope is not long enough for the weight to reach the ground. How high from the ground does the weight hang?

First we will find the location of the weight geometrically.

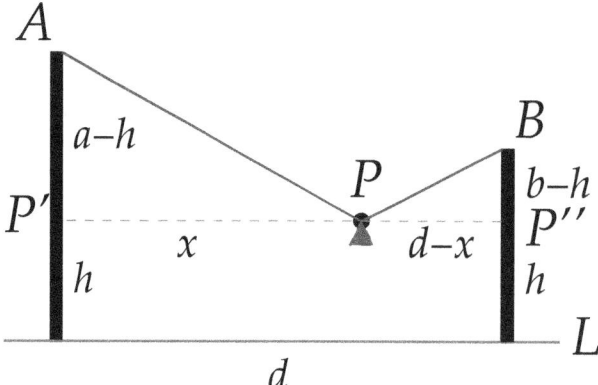

Calculation. *Let the height of the weight above the ground be h and let the distance from the weight to the left pole be x. Using the Pythagorean theorem, we get* $\sqrt{(a-h)^2+x^2}+\sqrt{(b-h)^2+(d-x)^2}=l$.

Recall from physics that $\angle APP'=\angle BPP''$. *Therefore triangles* APP' *and* BPP'' *are similar, so* $\dfrac{a-h}{x}=\dfrac{b-h}{d-x}$.

Thus we have a system of two equations with two unknowns. Although it is possible to solve this system, it is not easy. There is a nicer way to solve this problem.

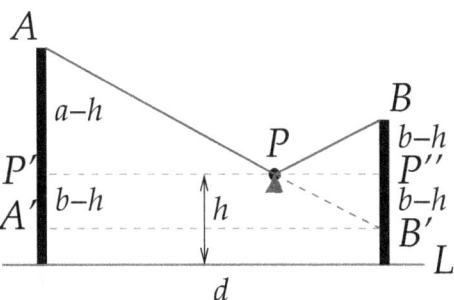

Construction. *Let P be the position of the weight which we have to find. Reflect B about the horizontal line* $P'P''$ *through P, let* B' *denote its image. Then* $PB=PB'$, *thus* $AB'=AP+PB'=AP+PB=l$. *Therefore to find* B', *we have to draw a circle of radius l centered at A, and then* B' *is its intersection point with the right pole. Once* B' *is found, divide* BB' *into two equal intervals with* P'' *being the midpoint. Draw a horizontal line through* P''. *Its intersection point with AB is the point P.*

To find h, let $A'B'$ *be the horizontal line through* B'. *Consider triangle* $A'B'A$. *Since* $A'B'=d$, $A'A=a+b-2h$, *and* $AB'=l$, *we have* $d^2+(a+b-2h)^2=l^2$. *Therefore* $a+b-2h=\sqrt{l^2-d^2}$, *thus* $2h=a+b-\sqrt{l^2-d^2}$, *and* $h=\dfrac{a+b-\sqrt{l^2-d^2}}{2}$.

In further examples we will only describe geometric constructions. We will not calculate positions of the objects algebraically.

Example 12.3. Two circles C and D, and a distance l are given. Construct a horizontal [1] segment XY of length l such that X lies on C and Y lies on D. Assume that such a segment exists.

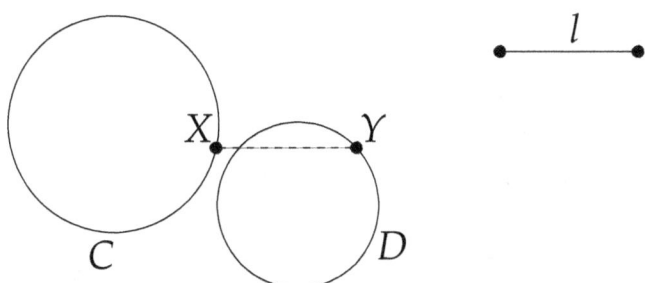

Solution. *Translate the circle C by the distance l to the right (or to the left, depending on where the second circle is). Let's call this new circle* C'. *Let Y be an intersection point of* C'

---

[1]We assume that a reference horizontal line is given. Thus any line parallel to it is horizontal and any line perpendicular to it is vertical.

and D if it exists. Draw a horizontal line through Y. Let X be the point on the horizontal line through Y, located a distance l to the left of Y. It lies on circle C.

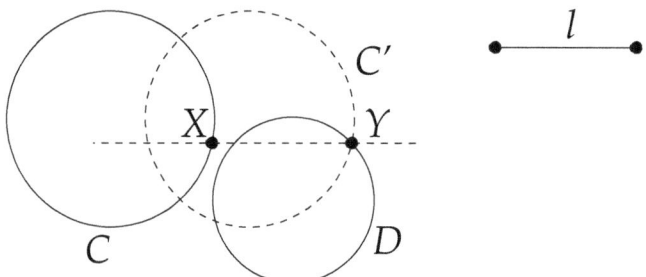

Note. If circles C and D are positioned so that C' and D do not intersect, then translate C to the left instead of to the right.

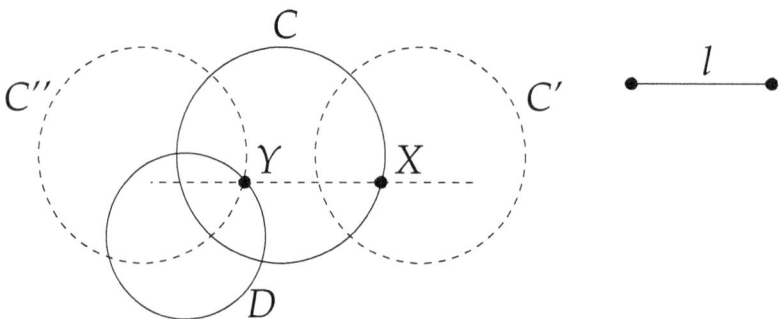

Example 12.4. Two distinct lines p and q, and a point S are given. Construct a square $ABCD$ that satisfies the following conditions (assume that such a square exists):

- Point S is the center of the square.

- Vertex A of the square lies on line p.

- Vertex B, the counterclockwise neighbor of A, lies on line q.

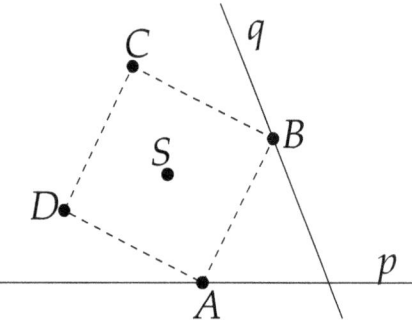

Solution. *Notice that segments SA and SB must be perpendicular and of the same length. Let's rotate line p through an angle of 90° in the counterclockwise direction around point S. Let p' be the new line. Let B be the intersection point of p' and q. Once we have one vertex and the center of the square, the rest is easy: draw line BS, find D such that SD = SB.*

Draw the line through S perpendicular to SB, and find A and C such that $SA = SC = SB$ with A lying on line p.

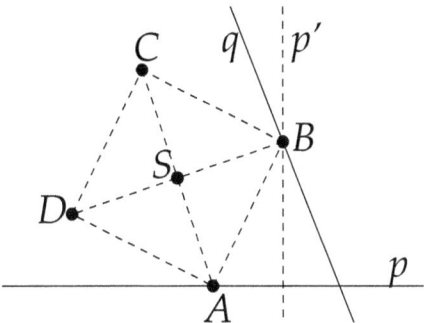

Problems

In all problems below, the words "find" and "draw" mean construct using a ruler and a compass (or using a geometry software as discussed above). You do not have to calculate the locations of all the points. Assume that solutions exist. Also as in example 12.3, assume that a reference horizontal line is given.

1. Two circles, S and T, and a point A are given. Find points B on S and C on T such that $\triangle ABC$ is isosceles with $|AB| = |AC|$, $\angle ABC = \angle ACB = 75°$, and $\angle BAC = 30°$.

2. Two distinct lines p and q and a point S are given. Construct a square $ABCD$ that satisfies the following conditions:

   - Point S is the center of the square.
   - Vertex A of the square lies on line p.
   - Vertex C, the opposite of vertex of A, lies on line q.

3. (a) Two points A and C and a line are given. Find a point B on the line such that $|AB| + |BC|$ is as small as possible.

    (b) A circle, a line, and a point C are given. The circle does not intersect the line. Find a point A on the circle and a point B on the line such that $|AB| + |BC|$ is as small as possible.

    (c) Two circles and a line are given. None of them intersect. Find a point A on one circle, a point B on the line, and a point C on the second circle such that $|AB| + |BC|$ is as small as possible.

4. A circle, a line, and a distance l are given. Find a point X on the circle and a point Y on the line such that the segment XY is horizontal and has length l.

5. A point A and two lines p and q are given. Find a point B on p and a point C on q such that $\triangle ABC$ is isosceles with $|AB| = |AC|$, and base BC is horizontal.

6. Line segments of lengths a, b, and c are given. Construct a line segment of length $\dfrac{ab}{c}$.

7. Line segments of lengths a, b, and c are given. Find a point on the segment of length a that divides it in the ratio $b : c$.

8. Line segments of lengths a and b are given. Construct a line segment of length $\sqrt{ab}$.

9. Two nonintersecting circles are given. Draw a line that is tangent to both circles and such that the circles lie

   (a) on one side of the line.
   (b) on opposite sides of the line.

10. (a) Show that among all rectangles with given perimeter, a square has the maximal area.

    (b) A farmer has 2400 ft of fencing and wants to fence off a rectangular field that borders a straight river. He needs no fence along the river. What are the dimensions of the field that has the largest area?

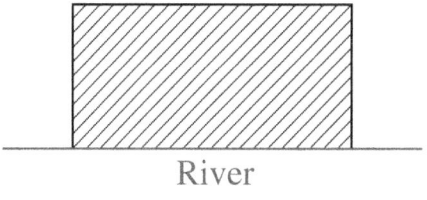

11. A length l and lines p and q are given. The lines intersect at a point C at an angle of $30°$. Line q is horizontal. Find points A on p and B on q such that

    • AB is vertical and A lies above q

    • The length of the bisector AD of $\angle CAB$ is equal to l.

12. Two lines and a point are given. The point does not lie on either of the lines. Draw a circle that is tangent to the given lines and passes through the given point.

13. Two lines p and q and a point A are given. Find points B on p and C on q such that $\triangle ABC$ is isosceles with $|AB| = |BC|$, and $\angle ABC = 90°$.

14. Two lines, p and q, and a point A are given. Construct a square $ABCD$ that satisfies the following conditions:

    • Vertex A is the given point.
    • Vertex B, the counterclockwise neighbor of vertex A, lies on line p.
    • Vertex C, the counterclockwise neighbor of vertex B, lies on line q.

15. The shaded region in the figure below is bounded by three semi-circles. Cut this region into four congruent parts, i.e. parts of equal size and shape.

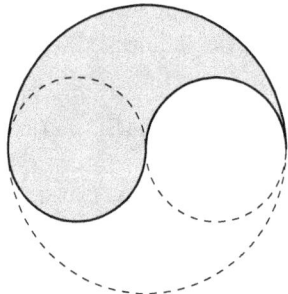

16. Three lines are given. Find three points on these lines, one point on each line, that are vertices of an equilateral triangle.

17. A point A and two lines, p and q, are given. Find a point B on line p, and a point C on line q, such that the perimeter of the triangle ABC is as small as possible.

18. Two circles with centers A and D are given. Find a point B on the first circle and a point C on the second circle, such that AB is horizontal and $|AB| + |BC| + |CD|$ is as small as possible.

19. Four lines, p, q, r, and s, and a distance l are given. Construct a horizontal line that intersects these four lines at points A, B, C, and D respectively in this order, and such that $|AB| + |CD| = l$.

20. Two spheres and a line are given in three-dimensional space. None of them intersect and neither of the spheres lies inside the other. Find a point A on one sphere, a point B on the line, and a point C on the other sphere so that $|AB| + |BC|$ is as small as possible.

Chapter 13

Graphs

Definition 13.1. A **graph** is an object consisting of a set of points called **vertices**, some of which are connected by lines (or arcs) called **edges**.

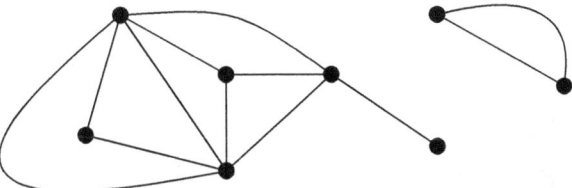

Definition 13.2. A graph is **simple** if any 2 vertices are connected by at most one edge and there are no loops (edges starting and ending at the same vertex).

Definition 13.3. If the edges are oriented, then we have an **oriented** or **directed** graph. An example of an oriented graph is a one-way road system.

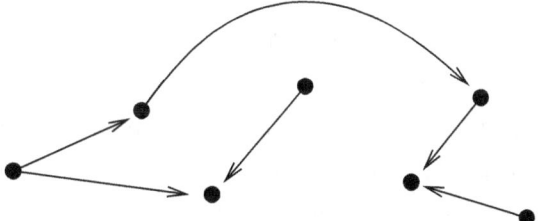

Definition 13.4. If an edge e connects the vertices v_1 and v_2, then we say that v_1 and v_2 are the **endpoints** of e and we write $e = (v_1, v_2)$. Also, we say that v_1 and v_2 are **adjacent vertices**. If two edges e_1 and e_2 share a common vertex, then we say that e_1 and e_2 are **adjacent edges**. A vertex v has **degree** m if m endpoints of edges coincide with v (a loop contributes 2 to the degree of a vertex).

Theorem 13.5. *In any graph, the sum of the degrees of the vertices equals twice the number of the edges.*

Corollary 13.6. *In any graph, the number of vertices with odd degrees is even.*

Definition 13.7. An undirected graph in which every two distinct vertices are connected by exactly one edge is called a **complete** graph. K_n denotes the complete graph with n vertices. The graphs K_2, K_3, K_4, and K_5 are shown below.

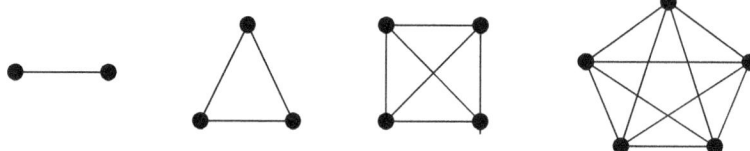

Definition 13.8. If the vertices of a graph can be separated into two parts X and Y so that for every edge in the graph, one of its endpoints belongs to X and the other belongs to Y, then we call such a graph a **bipartite** graph.

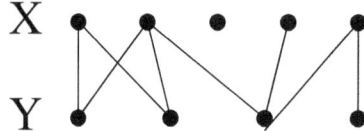

Definition 13.9. If every vertex in the set X is connected to every vertex in the set Y by exactly one edge, then the graph is called a **complete bipartite** graph. $K_{m,n}$ denotes the complete bipartite graph with m vertices in the set X and n vertices in the set Y. The graphs $K_{2,4}$ and $K_{3,3}$ are shown below.

Definition 13.10. We say that a graph can be **embedded** into a plane if it is possible to draw it in such a way that no two edges intersect. For example, the graph K_4 can be embedded as follows:

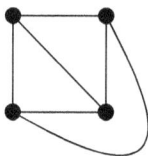

Theorem 13.11. *The graphs K_5 and $K_{3,3}$ can not be embedded into a plane.*

Definition 13.12. A **path** is a sequence of edges e_1, e_2, ..., e_n such that $e_1 = (x_0, x_1)$, $e_2 = (x_1, x_2)$, ..., $e_n = (x_{n-1}, x_n)$. When there are no multiple edges connecting the same vertices in a graph, a path can be denoted by its vertex sequence x_0, x_1, ..., x_n. A path that begins and ends at the same vertex is called a **cycle**. A path is **simple** if it does not contain the same edge more than once.

Definition 13.13. An **Euler path (resp. Euler cycle)** is a simple path (resp. cycle) containing every edge of the graph.

Theorem 13.14. *A connected graph has an Euler cycle if and only if each of its vertices has even degree.*

Definition 13.15. A **Hamilton path (resp. Hamilton cycle)** is a simple path (resp. cycle) visiting every vertex exactly once. (Note: in a cycle, the first and last vertices must coincide, so the start and finish count as one "visit".)

Definition 13.16. If all vertices of a graph can be visited by walking on edges, the graph is **connected**.

Definition 13.17. A connected graph without simple cycles is called a **tree**. Here is an example of a tree:

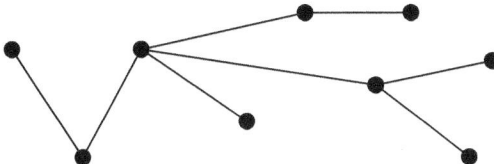

Example 13.18. Assume that in a group of people, any two either both know each other or both do not know each other. Prove that in any subset cointaining six people either there are three people that all know each other or there are three people that all do not know each other.

Solution. *Let's translate this problem into a graph theory problem. Let six vertices a, b, c, d, e, and f represent the six people. If two people know each other, then we use a red edge to join these two vertices. If two people do not know each other, then we use a blue edge to join these two vertices. Since there are edges between every two vertices in the graph, it's a complete graph K_6 with red and/or blue edges. Now the problem has been translated into the following problem: We use red and/or blue colors to color the edges in the complete graph K_6. Prove that there must exist either 3 vertices such that the edges joining them are all red, or 3 vertices such that the edges joining them are all blue. Now, let's pick any vertex in K_6, say a. The 5 edges between this vertex and the other 5 vertices are each either red or blue. According to Dirichlet's Principle, at least 3 edges of the five have the same color. Let's assume that (a,b), (a,c), (a,d) are red (the blue case is similar). Now consider the triangle bcd. If one of the edges (b,c), (b,d), (c,d) is red, then we have a red triangle. Otherwise, if (b,c), (b,d), (c,d) are all blue, then the triangle bcd is a blue triangle. This proves that there must exist a triangle all of whose edges are colored with the same color.*

Example 13.19. Is it possible to draw a triangular map inside a pentagon so that the degree of each vertex is even?

Below is an example of a triangular map (but some vertices have an odd degree):

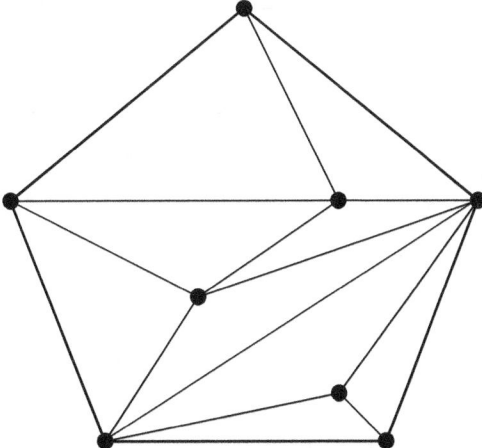

Solution. *The answer is no. We will prove this by contradiction. Suppose such a map exists. We know (see problem 9 in chapter 4) that every map with all vertices of even degree admits a proper coloring, i.e. its regions can be colored with 2 colors so that no neighboring regions have the same color. Color our map in blue and red so that the (infinite) region outside of*

the pentagon is blue. All the other regions are triangles. Each edge has a red triangle on one side and a blue region (either a triangle or that infinite outside region) on the other side. Now, count the number of edges (boundaries) in the map in two ways: each red triangle has 3 sides, so the number of edges is a multiple of 3, say, $3n$. Each blue triangle has 3 sides, and the infinite region has 5 edges, so the number of edges is a multiple of 3 plus 5, say, $3m + 5$. Thus we have $3n = 3m + 5$ where n and m are integers. But this is impossible.

Problems

1. Explain why a graph can not have 7 vertices of degrees 4, 4, 3, 3, 3, 2, 2.

2. (a) Does there exist a graph with 4 vertices of degrees 1, 1, 1, and 5?

   (b) Does there exist a simple graph with 4 vertices of degrees 1, 1, 1, and 5?

3. (a) Does there exist a graph with 8 vertices of degrees 7, 6, 5, 4, 3, 2, 1, 0?

   (b) Does there exist a simple graph with 8 vertices of degrees 7, 6, 5, 4, 3, 2, 1, 0?

4. (a) Does there exist a graph with 6 vertices of degrees 1, 2, 3, 4, 5, 5?

   (b) Does there exist a simple graph with 6 vertices of degrees 1, 2, 3, 4, 5, 5?

5. Prove that in any group of people, the number of people that are mutual friends with an odd number of people is even.

6. How many edges does a graph have if it has vertices of degrees 4, 3, 3, 2, 2? Draw such a graph.

7. Determine which of the following graphs are bipartite:

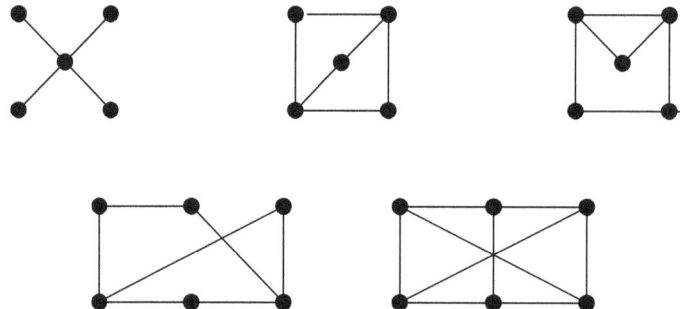

8. Which of the graphs in problem 7 have

   (a) an Euler path?

   (b) an Euler cycle?

   (c) a Hamilton path?

   (d) a Hamilton cycle?

9. There are 8 counties in Sikinia. There are no "four corners" points (like Arizona, Colorado, New Mexico, and Utah). Each county counted the number of neighboring counties. The numbers are 5, 5, 4, 4, 4, 4, 4, 3. Prove that at least one county made a mistake.

10. Find the number of vertices and edges in K_n and $K_{n,m}$.

11. Find a necessary and sufficient condition for a graph to have an Euler path but not an Euler cycle.

12. For what values of n does K_n have

    (a) an Euler path?

    (b) a Hamilton path?

13. For what values of n and m does $K_{n,m}$ have

    (a) an Euler cycle?

    (b) an Euler path?

    (c) a Hamilton cycle?

    (d) a Hamilton path?

14. A knight's tour is a sequence of legal moves by a knight starting at some square of a chessboard and visiting each square exactly once. (See problem 8 in chapter 1 for a description and picture of legal knight moves.) A knight's tour is called reentrant if there is a legal move that takes the knight from the last square of the tour back to where the tour began.

    (a) Draw the graph that represents all legal moves of a knight on a 3×4 chessboard.

    (b) Show that there is no reentrant tour on a 3×4 chessboard.

    (c) Find a non-reentrant tour on a 3×4 chessboard.

15. Show that there is no reentrant knight's tour on a 4×4 chessboard.

16. Show that there is no knight's tour at all (reentrant or not) on a 4×4 chessboard.

17. There are 7 men and 7 women attending a dance. After the dance, they recall the number of people they have danced with. The numbers are as follows: 3, 3, 3, 3, 3, 3, 3, 5, 6, 6, 6, 6, 6, 6. Prove that at least one of them made a mistake. (Assume that men only danced with women, and women only danced with men.)

18. There are 10 men and 10 women at a dance. Every man knows exactly 2 women and every woman knows exactly 2 men (assume that the relationship "knows" is mutual, i.e. either both people know each other or both people do not know each other). Prove that there exists a pairing such that every person would dance with one he/she knows.

19. There are 17 scientists who communicate with each other discussing some problems. They discuss only three topics, and each pair discusses at least one of these three. Prove that there are at least 3 scientists who are all pairwise discussing the same topic.

20. Nine mathematicians met at an international conference. They found that among any 3 of them there are at least 2 that have a language in common. If every mathematician speaks at most 3 languages, prove that at least three of the mathematicians can speak the same language.

21. Hamilton's "Round the World" puzzle: does the dodecahedron (shown below) have

    (a) a Hamilton path?

    (b) a Hamilton cycle?

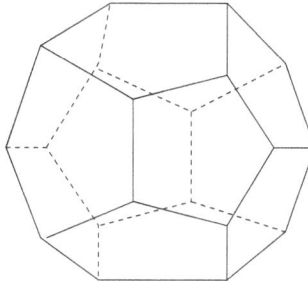

22. (a) Prove that in a finite simple graph having at least 2 vertices there are always two vertices with the same degree.

    (b) Does the above hold for graphs with loops but no multiple edges?

    (c) Does the above hold for graphs with multiple edges but no loops?

23. A connected bipartite graph G has 8 vertices. Recall that the vertices of a bipartite graph can be divided into 2 groups A and B so that every edge connects a vertex in group A and a vertex in group B. Both groups for G have 4 vertices. Three of the vertices in group A have degrees 4, 2, and 2. Three of the vertices in B have degrees 3, 1, and 1. What are the degrees of the remaining vertices?

24. A graph $K_{k,l,m}$ has $k + l + m$ vertices divided into three sets: k vertices in one set, l vertices in another set, and m vertices in the third set. Two vertices are connected if and only if they are in different sets. Prove that there do not exist $k, l, m \geq 1$ such that $K_{k,l,m}$ has exactly 6 edges.

25. Does $K_{1,2,4}$ have

    (a) an Euler cycle?

    (b) a Hamilton cycle?

    (c) a Hamilton path?

26. Find a necessary and sufficient condition on k, l, m for which $K_{k,l,m}$ has a Hamilton

    (a) cycle.

    (b) path.

27. Below is a map of the river and the bridges in Konigsberg. As we know from problem 7 in chapter 1, it is not possible to design a tour of the town that crosses each bridge exactly once and returns to the starting point. Could the citizens of Konigsberg create such a tour by building a new bridge?

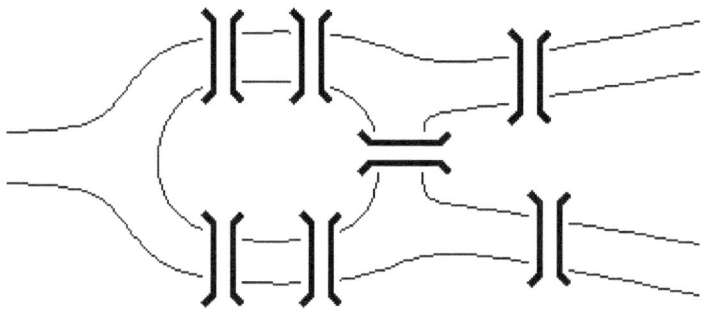

Chapter 14

Working backwards

"Working backwards" is a very powerful tool that can be used to solve many different problems.

Theorem 14.1. *Let a and b be nonzero integers, and let $d = (a, b)$. Then there exist integer numbers x and y such that $x \cdot a + y \cdot b = d$. When we find such integers a and b, we say that we wrote d as a linear combination of a and b.*

Euclid's algorithm. Given integers a and b, where $b \neq 0$, we can divide a by b and obtain quotient q and remainder r. Notice that since $a = qb + r$, the greatest common divisor of a and b is equal to the greatest common divisor of b and r. Euclid's algorithm of finding (a, b) and expression it as a linear combination of a and b is based on this fact:

$$
\begin{array}{llll|l}
a = q_1 \cdot b + r_1, & r_1 < b, & (a, b) = (b, r_1) & \quad & r_1 = a - q_1 \cdot b \\
b = q_2 \cdot r_1 + r_2, & r_2 < r_1, & (b, r_1) = (r_1, r_2) & & r_2 = b - q_2 \cdot r_1 \\
r_1 = q_3 \cdot r_2 + r_3, & r_3 < r_2, & (r_1, r_2) = (r_2, r_3) & & r_3 = r_1 - q_3 \cdot r_2 \\
\dots \downarrow & \dots & \dots & & \dots \uparrow \\
r_{n-2} = q_n \cdot r_{n-1} + r_n, & r_n < r_{n-1}, & (r_{n-2}, r_{n-1}) = (r_{n-1}, r_n) & & r_n = r_{n-2} - q_n \cdot r_{n-1} \\
r_{n-1} = q_{n+1} \cdot r_n, & \text{rem.} = 0, & (r_{n-1}, r_n) = r_n & &
\end{array}
$$

Thus $(a, b) = r_n$. Now we can work backwards using the equation on the right to express r_n as a linear combination of a and b.

Example 14.2. Find the greatest common divisor d of $a = 115$ and $b = 80$, and find x and y such that $x \cdot a + y \cdot b = d$.

Solution. *First perform division to find the gcd.*
$\quad 115 = 1 \cdot 80 + 35$
$\quad 80 = 2 \cdot 35 + 10$
$\quad 35 = 3 \cdot 10 + 5$
$\quad 10 = 2 \cdot 5$
$\quad$ *Therefore $(a, b) = 5$.*

Note. To find (a, b), we could factor $115 = 5 \cdot 23$, $80 = 2^4 \cdot 5$, so $(115, 80) = 5$. However, we need the above divisions to find the desired linear combination:

$\quad 5 = 35 - 3 \cdot 10,$
$10 = 80 - 2 \cdot 35,$
$35 = 115 - 1 \cdot 80,$
so

$$\begin{aligned}
5 &= 35 - 3 \cdot 10 \\
&= 35 - 3(80 - 2 \cdot 35) = 35 - 3 \cdot 80 + 6 \cdot 35 = 7 \cdot 35 - 3 \cdot 80 \\
&= 7(115 - 1 \cdot 80) - 3 \cdot 80 = 7 \cdot 115 - 7 \cdot 80 - 3 \cdot 80 = 7 \cdot 115 - 10 \cdot 80
\end{aligned}$$

Thus $x = 7$ and $y = -10$.

Example 14.3. Find a "closed" (i.e. not piece-wise defined) formula for the function whose graph is shown below.

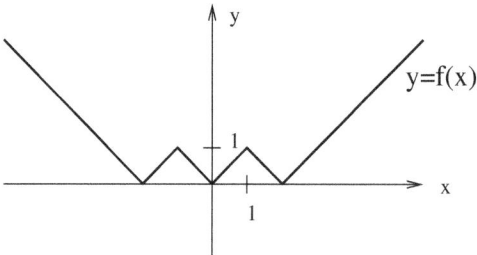

Solution. *Notice that $f(x)$ is the absolute value of the function $g(x)$ whose graph is*

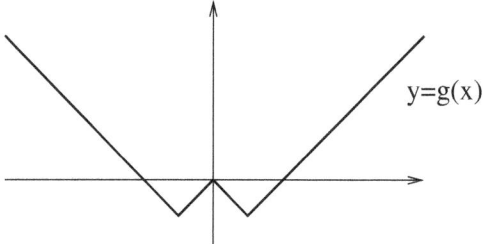

Record this fact: $f(x) = |g(x)|$. Here is the graph of $g(x) + 1$:

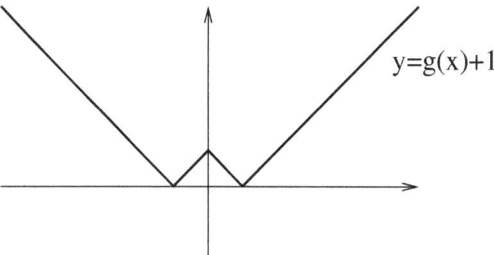

Notice that $g(x) + 1$ is the absolute value of $h(x)$ whose graph is

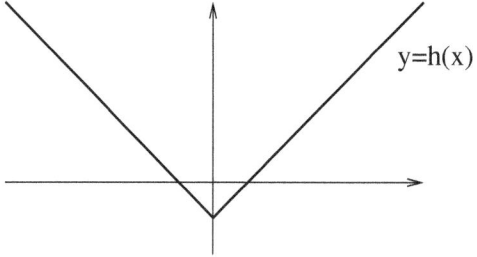

So, $g(x) + 1 = |h(x)|$. Finally, the graph of $h(x)$ is obtained from the graph of the absolute value of x by shifting it downward a distance of 1 unit, so $h(x) = |x| - 1$. Now, $g(x) + 1 = |h(x)| = ||x| - 1|$, so $g(x) = ||x| - 1| - 1$, and $f(x) = |g(x)| = |||x| - 1| - 1|$.

Example 14.4. Suppose 4 ones and 5 zeros are written along a circle. Between two equal numbers we write a one and between two distinct numbers we write a zero. Then the original numbers are wiped out. This step is repeated. Show that we can never reach 9 ones. For example, a possible initial distribution of ones and zeros and the first step are shown below:

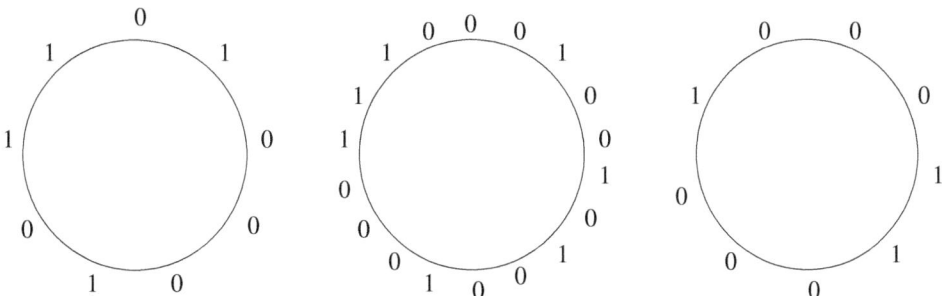

Solution. *Suppose the aim is attainable. Look at the first time we have 9 ones. One step before we must have 9 equal numbers. Since it was the first time we got 9 ones, one step before we must have 9 zeros. Still one step before we have 9 changes $0 - 1 - 0 - 1 - \ldots$. With an odd number of integers (9), this is not possible.*

Problems

1. Use Euclid's algorithm to find the greatest common divisor d of the given numbers a and b, and numbers x and y such that $xa + yb = d$.

   (a) $a = 46$, $b = 32$
   (b) $a = 24$, $b = 10$
   (c) $a = 96$, $b = 54$
   (d) $a = 219$, $b = 51$

2. Find integer numbers a and b such that $6 = 67a + 25b$.

3. Find a and b such that in Euclid's algorithm $r_7 = (a, b)$. Write out all the divisions.

4. Find $a > 100$ and $b > 100$ such that it will take exactly 5 divisions to reach the greatest common divisor of a and b.

5. Find positive a and b such that it will take at least 5 divisions to reach the greatest common divisor of a and b with the smallest possible number a and $b < a$.

In problems 6-10, find a closed formula for the function whose graph is shown.

6. A piece-wise defined function whose graph consists of arcs of cosine curves:

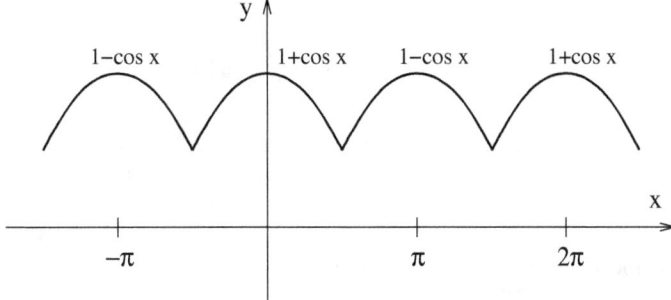

7. A piece-wise defined function shown below (the graph passes through $(-1, 3)$ and $(1, 3)$):

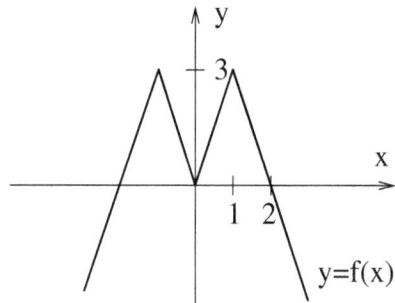

8. The following function has a value of 0 at all negative points x.

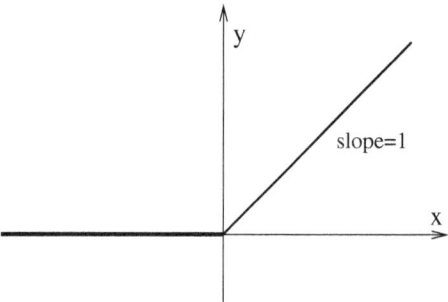

Hint: try subtracting $\dfrac{x}{2}$ from the given function.

9. This function also has a value of 0 at all negative points and the graph passes through $(1, 1)$.

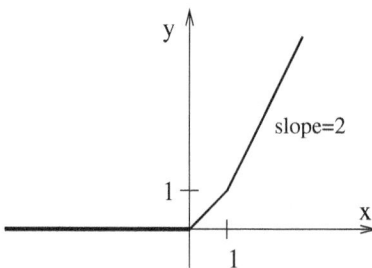

Hint: do problem (c) first.

10. A piece-wise defined function whose graph passes through $(-2, 0)$, $(0, 2)$, $(1, 1)$ and has slopes as indicated.

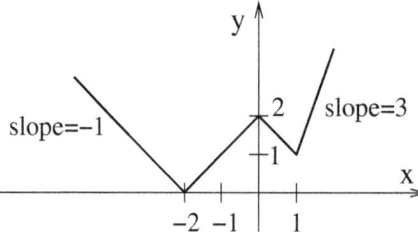

11. Starting with 2, 0, 0, 3, we construct the sequence 2, 0, 0, 3, 5, 8, 6, ..., where each new digit is the mod 10 sum of the preceding four terms. Prove that the 4-tuple 0, 5, 0, 5 will never occur.

12. Starting with 2, 0, 0, 3, we construct the sequence 2, 0, 0, 3, 5, 8, 6, ..., where each new digit is the mod 10 sum of the preceding four terms. Will the 4-tuple 0, 4, 0, 7 ever occur?

13. Two players play the following game. There are initially 27 counters in a pile. The players take turns removing 1, 2, 3, or 4 counters. The game ends when all counters have been removed. The player who takes the last counter loses. Find a winning strategy for one of the players.

14. Two players play the following game. There are initially 27 counters in a pile. The players take turns removing 1, 2, 3, or 4 counters. The game ends when all counters have been removed. The player who takes the last counter wins. Find a winning strategy for one of the players.

15. Two players play the following game. There are initially 10 counters in a pile. The players take turns removing 1 or 2 counters. The game ends when all counters have been removed. The player who takes the last counter loses. Find a winning strategy for one of the players.

16. There are two piles of candy. One pile contains 20 pieces, and the other 21. Two players take turns eating all the candy in one pile and separating the remaining candy into two (not necessarily equal) piles. (A pile may have 0 candies in it.) The player who cannot eat a candy on his/her turn loses. Which player, if either, can guarantee victory in this game?

17. Two players play the following game. There are initially 10 counters in a pile. The players take turns removing 1, 2, or 4 counters. The game ends when all counters have been removed. The player who takes the last counter wins. Find a winning strategy for one of the players.

18. Two players play the following game. Initially $X = 0$. The players take turns adding any number between 1 and 10 (inclusive) to X. The game ends when X reaches 100. The player who reaches 100 wins. Find a winning strategy for one of the players.

19. Two players play the following game. There are initially 50 counters in a pile. The players take turns removing 1, 2, 4, 8, 16, or 32 counters. The game ends when all counters have been removed. The player who takes the last counter wins. Find a winning strategy for one of the players.

20. In problem 19, what is the smallest number N such that no matter how the second player plays, the first player can win in N or less moves?

21. A box contains 300 matches. Two players take turns removing no more than half the matches in the box. The player who cannot take any match(es) loses. Find a winning strategy for one of the players.

22. Two players play the following game. Initially $X = 1$. The players take turns multiplying X by any whole number from 2 to 9 (inclusive). The player who first names a number greater than 1000 wins. Which player, if either, can guarantee victory in this game?

23. Suppose you are writing a calculus book. You want to find a few cubic polynomials $f(x) = ax^3 + bx^2 + cx + d$ (preferably with integer coefficients) whose critical numbers are integers. (Recall that a critical number is a value of x at which the derivative is

equal to 0.) How would you find such polynomials? Use your strategy to find a couple of polynomials.

24. Suppose you want to give your high school students a system of 2 linear equations with 2 variables. You'd like the answers to be integer numbers. You could, of course, try random coefficients, say

$$\begin{cases} 2x + 3y = 4 \\ 5x - 6y = 7 \end{cases},$$

solve your systems, and hope that sooner or later you'll find a system with integer solutions, but is there a better strategy?

25. Suppose you are teaching linear algebra, and you need to find matrices with integer entries whose reduced echelon forms also have integer entries. How would you find such matrices?

26. The integers 1, 2, ..., n are placed in order, so that each value is either bigger than all preceding values or is smaller than all preceding values. In how many ways can this be done?

27. I have seven coins whose total value is \$0.57. What coins do I have? And, how many of each coin do I have? (Coins being used at the time when this book is written have values 1 cent, 5 cents, 10 cents, 25 cents, and 1 dollar.)

Chapter 15

Calculus

Recall the following important definitions and theorems.

Definition 15.1. If a, x, $y \in \mathbb{R}$ and $a > 0$, $a \neq 1$, $x > 0$, then

$$\log_a x = y \quad \Leftrightarrow \quad a^y = x$$

Theorem 15.2. *(Properties of logarithms)*

1. $\log_a(xy) = \log_a x + \log_a y$

2. $\log_a\left(\dfrac{x}{y}\right) = \log_a x - \log_a y$

3. $\log_a(x^r) = r \log_a x$

4. $\log_a(x) = \dfrac{\ln x}{\ln a}$

Definition 15.3. A function $f(x)$ is called **even** if $f(-x) = f(x)$ for all x in the domain of f.
A function $f(x)$ is called **odd** if $f(-x) = -f(x)$ for all x in the domain of f.

Definition 15.4. Let $f : A \to B$ be a function. A function $f^{-1} : B \to A$ is called the **inverse** of f if for all $x \in A$ and $y \in B$

$$f^{-1}(y) = x \quad \Leftrightarrow \quad f(x) = y.$$

Theorem 15.5. *If f^{-1} is the inverse of $f : \mathbb{R} \to \mathbb{R}$, then the graphs of $y = f(x)$ and $y = f^{-1}(x)$ are symmetric about the line $y = x$.*

Theorem 15.6. *(Intermediate value theorem) Suppose $f(x)$ is continuous on $[a, b]$. Let N be any number between $f(a)$ and $f(b)$. Then there exists $c \in [a, b]$ such that $f(c) = N$.*

Definition 15.7. The derivative of $f(x)$ at a point a is

$$f'(a) = \lim_{h \to 0} \frac{f(a + h) - f(a)}{h}.$$

The derivative $f'(a)$ is the slope of the tangent line to $y = f(x)$ at $(a, f(a))$. Also, $f'(a)$ is the rate of change of $f(x)$ with respect to x at $x = a$.

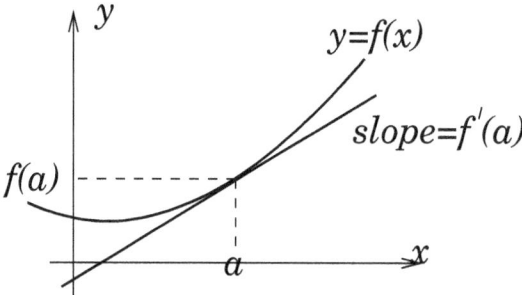

Theorem 15.8. *(Important derivatives)*

$$(x^n)' = nx^{n-1}, \qquad (e^x)' = e^x, \qquad (a^x)' = (\ln a)a^x,$$

$$(c)' = 0, \qquad (\ln x)' = \frac{1}{x}, \qquad (\log_a x)' = \frac{1}{(\ln a)x},$$

$$(\sin x)' = \cos x, \qquad (\cos x)' = -\sin x, \qquad (\tan x)' = (\sec x)^2,$$

$$(\csc x)' = -\csc x \cot x, \qquad (\sec x)' = \sec x \tan x, \qquad (\cot x)' = -(\csc x)^2,$$

$$(\arcsin x)' = \frac{1}{\sqrt{1-x^2}} \qquad (\arccos x)' = -\frac{1}{\sqrt{1-x^2}}, \qquad (\arctan x)' = \frac{1}{x^2+1}$$

Theorem 15.9. *(Chain rule)* $(f \circ g)'(x) = (f(g(x)))' = f'(g(x))g'(x)$

Theorem 15.10. *If $f(x)$ is defined on some open interval containing a point c and has a local maximum or minimum c, then c is a critical number of $f(x)$ (i.e. either $f'(c) = 0$ or $f'(c)$ does not exist).*

Definition 15.11. Let $f(x)$ be continuous on an interval $[a, b]$. Divide the interval into n subintervals of equal length: $[x_0, x_1]$, $[x_1, x_2]$, ..., $[x_{n-1}, x_n]$ where $x_0 = a$ and $x_n = b$. Let $\Delta x = \dfrac{b-a}{n}$ be the length of each subinterval. Then the sum

$$R_n = \sum_{i=1}^{n} f(x_i)\Delta x$$

is called the **Riemann sum** of $f[x]$ on $[a, b]$ using n subintervals. It can be proved that the limit of R_n as n approaches infinity exists, and

$$\int_a^b f(x)dx = \lim_{n \to \infty} \sum_{i=1}^{n} f(x_i)\Delta x$$

is called the integral of $f(x)$ from a to b.

If $f(x) \geq 0$, then $\displaystyle\int_a^b f(x)dx$ is the area of the region under the curve $y = f(x)$ and above the x-axis from a to b.

If $f(x)$ takes on both positive and negative values, then $\displaystyle\int_a^b f(x)dx$ is the sum of the areas under the curve and above the x-axis minus the sum of the areas under the x-axis and above the curve.

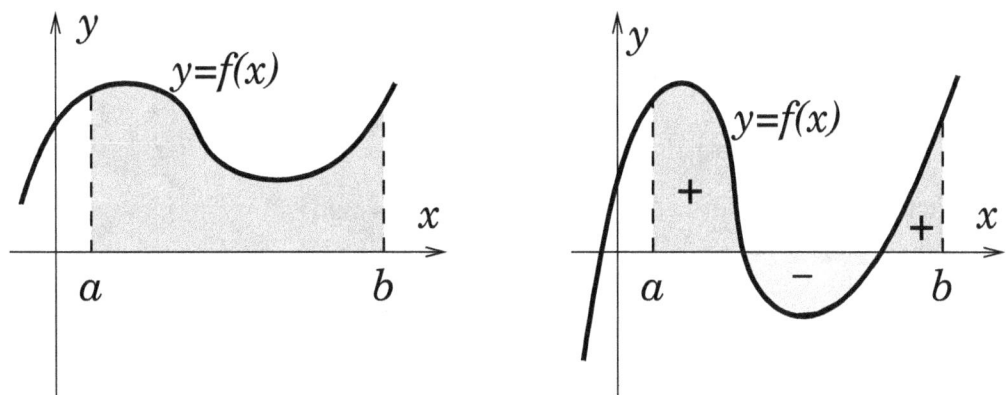

Theorem 15.12. *(Fundamental Theorem of Calculus)*

I. $\displaystyle \frac{d}{dx}\left(\int_a^x f(t)dt\right) = f(x)$

II. *If* $F'(x) = f(x)$, *then* $\displaystyle\int_a^b f(x)dx = F(b) - F(a)$.

Theorem 15.13. *(Substitution Rule)*

$\displaystyle\int f(g(x))g'(x)dx = \int f(u)du \quad where \quad u = g(x), \; du = g'(x)dx.$

Theorem 15.14. *(Some important series)*

$\displaystyle\sum_{n=1}^{\infty} \frac{1}{n} = 1 + \frac{1}{2} + \frac{1}{3} + \frac{1}{4} + \ldots \quad is \; divergent.$

$\displaystyle\sum_{n=0}^{\infty} q^n = 1 + q + q^2 + q^3 + \ldots = \begin{cases} \dfrac{1}{1-q} & if \quad |q| < 1, \\ divergent & if \quad |q| \geq 1. \end{cases}$

$\displaystyle\sum_{n=0}^{\infty} \frac{x^n}{n!} = 1 + \frac{x}{1!} + \frac{x^2}{2!} + \frac{x^3}{3!} + \ldots = e^x \quad for \; all \; x.$

(in particular, if $x = 1$, then $\displaystyle\sum_{n=0}^{\infty} \frac{1}{n!} = 1 + \frac{1}{1!} + \frac{1}{2!} + \frac{1}{3!} + \ldots = e.$)

$\displaystyle\sum_{n=0}^{\infty}(-1)^n \frac{x^{2n+1}}{2n+1} = x - \frac{x^3}{3} + \frac{x^5}{5} - \frac{x^7}{7} + \ldots = \arctan x \quad for \; all \; x.$

(in particular, if $x = 1$, then $\displaystyle\sum_{n=0}^{\infty}(-1)^n \frac{1}{2n+1} = 1 - \frac{1}{3} + \frac{1}{5} - \frac{1}{7} + \ldots = \arctan 1 = \frac{\pi}{4}.$)

Problems

1. Evaluate the integral $\int_{-4}^{2} |x+2|dx$.

2. Evaluate the integral $\int_{0}^{3\pi} |\sin x|dx$.

3. Find a number c such that the line $y = x - 1$ is tangent to the parabola $y = cx^2$.

4. (a) Show that the function $f(x) = \ln(x + \sqrt{x^2 + 1})$ is odd.

   (b) Find the inverse of $f(x)$.

5. Find a cubic polynomial $p(x) = ax^3 + bx^2 + cx + d$ that has a local maximum at $(0, 1)$ and a local minimum at $(1, 0)$.

6. Find the interval $[a, b]$ for which the value of the integral $\int_{a}^{b} (2+x-x^2)dx$ is a maximum.

7. Find all values of a for which the area of the region bounded by the line $y = ax$ and the parabola $y = x^2$ is equal to 1.

8. There is a line through the origin that divides the region bounded by the parabola $y = x - x^2$ and the x-axis into two regions with equal area. What is the slope of that line?

9. Find the sum of the series $\displaystyle\sum_{n=0}^{\infty} \frac{1}{2^{2n+1}} = \frac{1}{2} + \frac{1}{2^3} + \frac{1}{2^5} + \frac{1}{2^7} + \dots$

   Hint: e.g. factor out $\dfrac{1}{2}$, and notice that $2^{2n} = 4^n$.

10. Find the sum of the series
$$1 + \frac{1}{2} + \frac{1}{3} + \frac{1}{4} + \frac{1}{6} + \frac{1}{8} + \frac{1}{9} + \frac{1}{12} + \frac{1}{16} + \frac{1}{18} + \dots$$

   where the terms are the reciprocals of the positive integers whose only prime factors are 2s and 3s.

11. The parabola $y = x^2 + 2$ has two tangent lines that pass through the origin. Find their equations.

12. Suppose you have a large supply of books, all the same size, and you stack them at the edge of a table, with each book extending farther beyond the edge of the table than the one beneath it. Show that it is possible to do this so that the top book extends entirely beyond the table. In fact, show that the top book can extend any distance at all beyond the edge of the table if the stack is high enough. Try the following method of stacking: The top book extends half its length beyond the second book. The second book extends a quarter of its length beyond the third. The third extends one-sixth of its length beyond the fourth, and so on. (You could try it yourself with a deck of cards or CDs. To make your construction stable, make each book/card/CD extend a little bit less than described in this problem.) Hint: consider centers of mass.

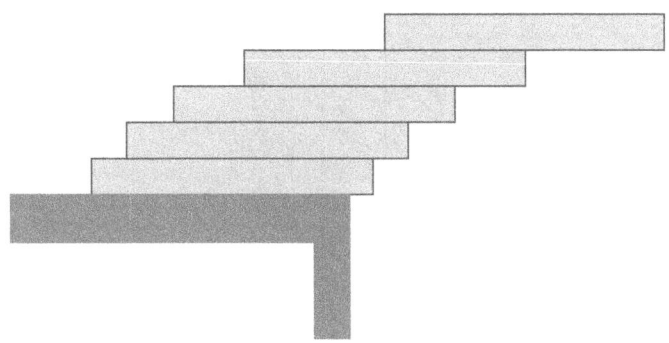

13. Find the n-th derivative of $f(x) = \dfrac{1}{x^2 + x}$.

Hint: use the partial fraction decomposition. Recall that since $x^2 + x = x(x + 1)$, the partial fraction decomposition has the form $\dfrac{A}{x} + \dfrac{B}{x + 1}$.

14. Find the n-th derivative of the function $f(x) = \dfrac{x^n}{1 - x}$.

15. The parabola $y = x^2$ and the line $y = mx + 1$ are given. They have two intersection points, A and B. Find the point C on the parabola that lies between A and B and maximizes the area of $\triangle ABC$.

16. The figure below shows a curve C with the property that, for every point P on the middle curve $y = 2x^2$, the areas A and B are equal. Find an equation for C.

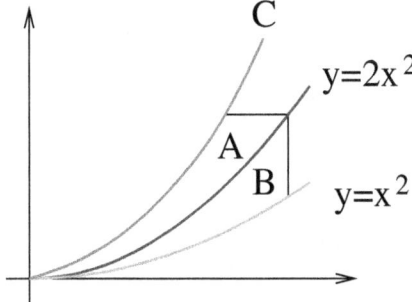

17. Find all values of a such that $x^2 + ax + 1 \geq \cos x$ for all real x.

18. For which positive numbers a is it true that $a^x \geq 1 + x$ for all x?

19. The figure below shows a circle with radius 1 inscribed in the parabola $y = x^2$. Find the center of the circle.

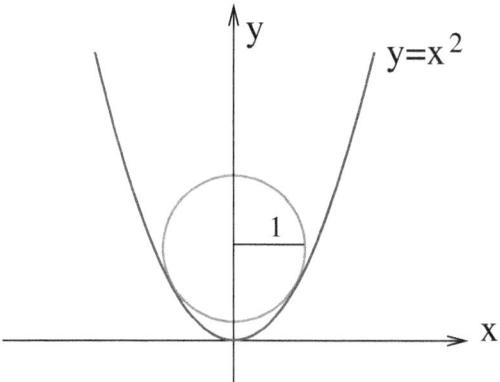

20. The figure below shows a region consisting of all points inside a unit square that are closer to the center than to the sides of the square. Find the area of the region.

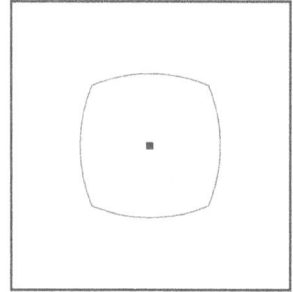

21. Sketch the region $S = \{(x, y) \mid |x| \geq 1, \; |y| \geq 2, \; x^2 + y^2 \leq 9\}$ and find its area.

22. Find a continuous function f such that the area under the graph of f from 0 to t is $A(t) = t^3$ for all $t > 0$.

23. The figure below shows a horizontal line $y = c$ intersecting the curve $y = 8x - 27x^3$. Find the number c such that the areas of the shaded regions are equal.

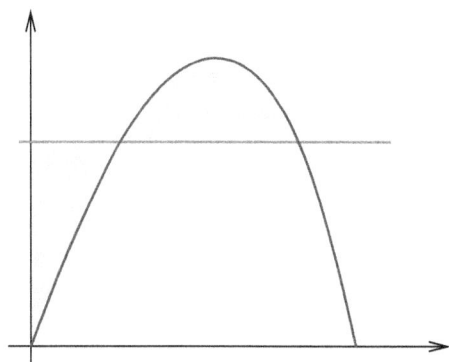

24. Evaluate $\displaystyle \lim_{n \to \infty} \left(\frac{1}{\sqrt{n}\sqrt{n+1}} + \frac{1}{\sqrt{n}\sqrt{n+2}} + \ldots + \frac{1}{\sqrt{n}\sqrt{n+n}} \right)$.
    Hint: interpret the sum as a Riemann sum of a function. Then the limit as n approaches infinity is the value of an integral.

25. Show that any ellipsoid (given by $\dfrac{x^2}{a^2} + \dfrac{y^2}{b^2} + \dfrac{z^2}{c^2} = 1$) has a section that is a circle.
    Hint: any section of the ellipsoid that passes through the origin is an ellipse.

26. Evaluate $\displaystyle\int \dfrac{1}{x^7 - x}\,dx$.
    The straightforward approach would be to start with partial fractions, but that would be too brutal. We could reduce the power of the denominator as follows:
    $\displaystyle\int \dfrac{1}{x^7 - x}\,dx = \int \dfrac{x}{x^8 - x^2}\,dx$, let $u = x^2$, then $du = 2x\,dx$, or $\dfrac{du}{2} = x\,dx$, and we have
    $\displaystyle\int \dfrac{x}{x^8 - x^2}\,dx = \dfrac{1}{2}\int \dfrac{1}{u^4 - u}\,du$.
    $u^4 - u$ is better than $x^7 - x$, but can you find an even better substitution?

27. Let $a_1, a_2, \ldots, a_{30}$ be real numbers. Show that $a_1\cos x + a_2\cos(2x) + \ldots + a_{30}\cos(30x)$ cannot take on only positive values.

28. If $a_0, a_1, a_2, \ldots, a_k$ are real numbers and $a_0 + a_1 + a_2 + \ldots + a_k = 0$, show that
$$\lim_{n\to\infty} (a_0\sqrt{n} + a_1\sqrt{n+1} + a_2\sqrt{n+2} + \ldots a_k\sqrt{n+k}) = 0.$$

    Hint: try the special cases $k = 1$ and $k = 2$ first, and then generalize.

29. Show that for $x > 0$,
$$\dfrac{x}{x^2 + 1} < \arctan x < x.$$

30. The figure below shows a point P on the parabola $y = x^2$ and the point Q where the perpendicular bisector of OP intersects the y-axis. As P approaches the origin along the parabola, what happens to Q? Does it have a limiting position? If so, find it.

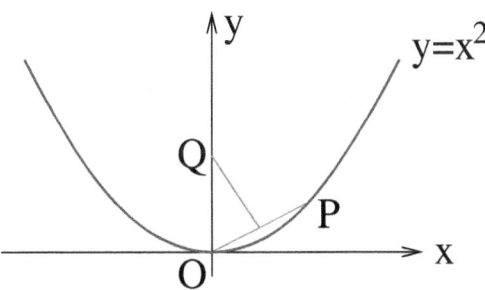

31. Recall that the area of a circle with radius r is $A = \pi r^2$ and the circumference of the circle is $L = 2\pi r$. Notice that $(\pi r^2)' = 2\pi r$. Similarly, the volume of a ball with radius r is $V = \dfrac{4}{3}\pi r^3$, the surface area is $S = 4\pi r^2$, and $\left(\dfrac{4}{3}\pi r^3\right)' = 4\pi r^2$. Is this a coincidence? Actually, it isn't. Explain these facts. What is the ratio of the 4-dimensional volume to the usual 3-dimensional volume of its boundary (the analog of the surface area) for a 4-dimensional ball with radius r?

32. What is the ratio of the 5-dimensional volume of a 5-dimensional ball of radius r to the 4-dimensional volume of its boundary (the analog of the surface area)?

33. Evaluate $\displaystyle\int_0^1 (\sqrt[3]{1 - x^7} - \sqrt[7]{1 - x^3})\,dx$.

34. Show that e is irrational.

35. Let $f(x) = a_1 \sin x + a_2 \sin(2x) + a_3 \sin(3x) + \ldots + a_n \sin(nx)$, where $a_1, \ldots, a_n$ are real numbers and n is a positive integer. If it is given that $|f(x)| \le |\sin(x)|$ for all x, show that $|a_1 + 2a_2 + \ldots + na_n| \le 1$.

36. Let $T(x)$ denote the temperature at the point x on Earth at some fixed time. Assuming that T is a continuous function of x, show that at any fixed time there are at least two diametrically opposite points on the equator that have the same temperature.

37. Find a curve that passes through the point $(3, 2)$ and has the property that if the tangent line is drawn at any point P on the curve, then the part of the tangent line that lies in the first quadrant is bisected by P.

38. Explain why the curve shown below cannot be the graph of a cubic polynomial.

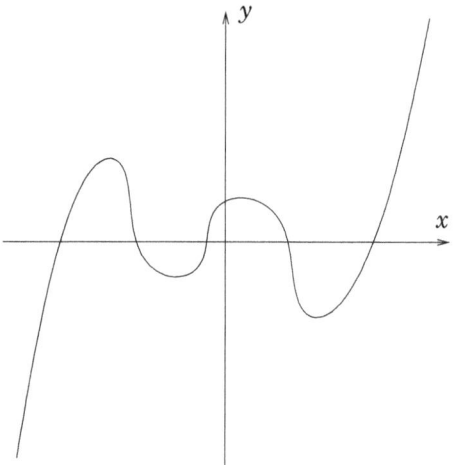

39. Evaluate the integral: $\displaystyle\int_0^1 \arcsin(x)\,dx$ (hint: use areas).

40. Find the volume of a 4-dimensional unit ball.

Chapter 16

Various problems

Most problems in this section can be solved in a few different ways.

Problems

1. Show that there is no reentrant knight's tour on a 5×5 chessboard (recall that a tour is called reentrant if there is a legal move that takes the knight from the last square of the tour back to where the tour began).

2. Prove that for any integer number n, $n^7 - n$ is divisible by 7.

3. A sequence $\{a_n\}$ is defined recursively by the equations
$$a_0 = a_1 = 1, \qquad n(n-1)a_n = (n-1)(n-2)a_{n-1} - (n-3)a_{n-2}.$$
Find the sum of the series $\sum_{n=0}^{\infty} a_n$.

4. Evaluate the integral: $\displaystyle\int_{-2}^{3} \left||x| - 1\right| dx$

5. Solve the inequality: $\left|6 - |x| - x\right| + x \leq 3$.

6. A 6×6 rectangle is tiled by 2×1 dominoes. Prove that it has at least one *fault-line*, that is, a straight line cutting the rectangle without cutting any domino.

7. (a) Find an example of a polygon and a point in its interior, so that no side of the polygon is completely visible from that point.

   (b) Find an example of a polygon and a point in its exterior, so that no side of the polygon is completely visible from that point.

8. The plane is colored with two colors. Prove that there exist three points of the same color, which are vertices of a regular triangle.

9. The plane is colored with n colors where n is any natural number. Prove that there exist four points of the same color, which are vertices of a rectangle.

10. Each block of a 25×25 board has either 1 or -1 written on it. Let a_i be the product of all numbers in the ith row and b_j be the product of all numbers in the jth column. Prove that $a_1 + \ldots + a_{25} + b_1 + \ldots + b_{25} \neq 0$.

11. December 25, 2005 was a Sunday. What day of the week is going to be December 25, 2025?

12. Which natural numbers are sums of consecutive smaller natural numbers? For example, $30 = 9 + 10 + 11$ and $31 = 15 + 16$, but 32 has no such representation. Find a simple necessary and sufficient condition.

13. Prove that if 40 coins are distributed among 9 bags so that each bag contains at least one coin, then at least two bags contain the same number of coins.

14. *The Art Gallery Problem.* An art gallery has the shape of an n-gon (not necessarily a convex one). The boundary of the n-gon are the only walls, there are no walls inside it. The gallery is to be guarded by security cameras that can each rotate to obtain a full field of vision. Prove that $\left[\dfrac{n}{3}\right]$ (the integer part of $\dfrac{n}{3}$) cameras can survey the building, no matter how complicated its shape.

For example:

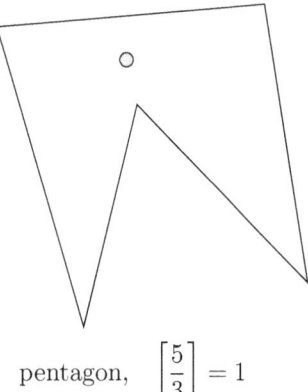

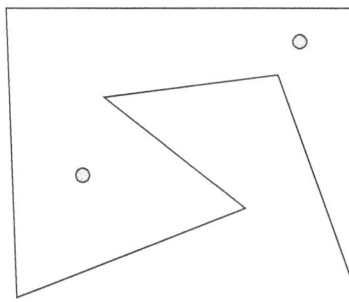

pentagon, $\left[\dfrac{5}{3}\right] = 1$ 7-gon, $\left[\dfrac{7}{3}\right] = 2$

1 camera can survey the building   2 cameras can survey the building

Note. There are many variations of this problem. In some versions cameras are restricted to the walls (edges of the polygon), or even to the corners (vertices of the polygon). Some versions require only the edges to be guarded. There is a whole book (Art Gallery Theorems and Algorithms, Joseph O'Rourke, Oxford University Press 1987) dedicated to this problem.

Chapter 17

Solutions and answers to selected problems

17.1   Introduction

1. Assume that each of the eleven children contributed at most $2.72. Then the total amount does not exceed $2.72 \cdot 11 = 29.92$ dollars. But the total amount is $30.00. Therefore our assumption is false, thus at least one child contributed at least $2.73. This kind of proof is called a proof by contradiction (see chapter 3). Such problems can also be solved using the Generalized Dirichlet's Principle (see chapter 5).

2. (a) Any two-digit number N can be written in the form $N = 10a + b$ where b is the units digit of the number and a is its tens digit. (For example, $27 = 10 \cdot 2 + 7$.)
    Suppose that N is divisible by 3. Then $N = 3k$ for some integer k. Thus
    $$10a + b = 3k$$
    $$9a + a + b = 3k$$
    $$a + b = 3k - 9a$$
    $$a + b = 3(k - 3a)$$
    Since $k - 3a$ is an integer, $a + b$ is divisible by 3.
    Conversely, suppose that $a + b$ is divisible by 3. Then $a + b = 3m$ for some integer m. Thus
    $$9a + a + b = 9a + 3m$$
    $$10a + b = 3(3a + m)$$
    $$N = 3(3a + m)$$
    Since $3a + m$ is an integer, N is divisible by 3.

    (b) Any natural number N can be written in the form

    $$N = 10^n a_n + 10^{n-1} a_{n-1} + \ldots 10a_1 + a_0$$

    where a_0 is the units digit of the number, a_1 is the tens digit, and so on (this is called the base 10 expansion of the number N, see chapter 6). Now,

    $$
    \begin{aligned}
    N &= \underbrace{99\ldots9}_{n} a_n + a_n + \underbrace{99\ldots9}_{n-1} a_{n-1} + a_{n-1} + \ldots + 9a_1 + a_1 + a_0 \\
    &= \underbrace{99\ldots9}_{n} a_n + \underbrace{99\ldots9}_{n-1} a_{n-1} + \ldots + 9a_1 + (a_n + a_{n-1} + \ldots + a_1 + a_0).
    \end{aligned}
    $$

Since all multiples of 9 are divisible by 3, an argument similar to the one in part (a) shows that N is divisible by 3 if and only if

$$a_n + a_{n-1} + \ldots a_1 + a_1$$

is (also, see chapter 6 for divisibility properties).

3. False. For example, for $n = 41$, $n^2 + n + 41 = 41^2 + 41 + 41 = 41 \cdot 43$ is not prime.

Note. You may be tempted to check a few small values of n. You would discover then that for $1 \leq n \leq 39$, the number $n^2 + n + 41$ is indeed prime. However, the above example shows that this is not the case for *all* natural values of n. Thus checking a few examples is not sufficient!

4. Choose a row with the biggest number of stars in it. Note that this row contains at least two stars since if each row contained at most one star then there would be at most 4 stars total. But there are 6 stars. (This argument is using Dirichlet's box principle, see chapter 5.) So we have the following three cases:

Case I. This row contains 4 stars. Then cross is out, and there will only be two stars left. If they are in different columns, then cross out any other row, and the two columns containing the remaining two stars. If the remaining two stars are in one column then cross out one more row, the column containing the stars, and any other column.

Case II. This row contains 3 stars. Then cross it out, and there will only be three stars left. Clearly we can eliminate three stars by crossing out one more row and two columns.

Case III. This row contains 2 stars. Cross it out, and there will only be four stars in three rows left. Therefore at least one of these rows contains two stars. Cross it out, and there will only be two stars left. As above, it is clear that we can eliminate two stars by crossing out two columns.

5. True. There are several ways to prove this. One is by induction (see chapter 4), another one is considering all possible remainders of n modulo 3 (see chapters 6 and 7). Here is a third way: $n^3 + 2n = n^3 - n + 3n = n(n^2 - 1) + 3n = n(n-1)(n+1) + 3n$. The term $n(n-1)(n+1)$ is the product of three consecutive numbers, and one of them is divisible by 3 (see chapter 6), therefore the product is divisible by 3. The term $3n$ is also divisible by 3. Thus the sum is divisible by 3.

6. Since $|x + 2| = \begin{cases} x + 2 & \text{if } x + 2 \geq 0, \ \text{i.e. if } x \geq -2 \\ -x - 2 & \text{if } x + 2 < 0, \ \text{i.e. if } x < -2 \end{cases}$ and

$|2x - 5| = \begin{cases} 2x - 5 & \text{if } 2x - 5 \geq 0, \ \text{i.e. if } x \geq 2.5 \\ -2x + 5 & \text{if } 2x - 5 < 0, \ \text{i.e. if } x < 2.5 \end{cases}$, we have

$$f(x) = |x + 2| + |2x - 5| = \begin{cases} x + 2 + 2x - 5 = 3x - 3 & \text{if } x \geq 2.5 \\ x + 2 - 2x + 5 = -x + 7 & \text{if } -2 \leq x < 2.5 \\ -x - 2 - 2x + 5 = -3x + 3 & \text{if } x < -2 \end{cases}$$

So we draw the graph of each linear function for the corresponding interval:

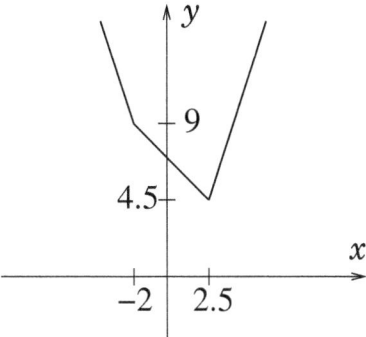

7. No. Consider the four regions of the town, namely the two river banks and the two islands. Each of them is connected with other regions by either 3 or 5 bridges. Suppose that there is a tour of the town that crosses every bridge exactly once. For each intermediate region on such a tour we must come to the region by a bridge and leave the region by a bridge. So every time the tour visits a region, two bridges are crossed. This means that for every region except the one where we start and the one where we end there must be an even number of bridges connecting that region to others. But we have 4 regions with an odd number of bridges. Thus we get a contradiction. This solution can be explained in an easier and "smoother" way if we use the graph terminology discussed in chapter 13.

8. Yes. There are even many such tours. Below are two tours (squares are numbered in the order the knight can visit them). Notice that in the second one, it is possible to go from square number 64 back to square number 1. Such a tour is called reentrant (see chapter 13).

1	40	13	26	3	42	15	28
24	37	2	41	14	27	4	43
39	12	25	60	53	62	29	16
36	23	38	63	56	59	44	5
11	50	57	54	61	52	17	30
22	35	64	51	58	55	6	45
49	10	33	20	47	8	31	18
34	21	48	9	32	19	46	7

1	14	17	42	3	38	19	40
16	43	2	63	18	41	4	37
13	64	15	58	53	56	39	20
44	27	12	55	62	59	36	5
11	30	61	52	57	54	21	50
26	45	28	31	60	51	6	35
29	10	47	24	33	8	49	22
46	25	32	9	48	23	34	7

Note. You may not be able to find such a tour quickly. However, it is important to recognize the fact that if you tried and were unable to find one, that does not prove that such a tour does not exist.

17.2   Logic

1. Construct the truth tables:

(a)

p	q	$p \rightarrow q$	$\neg q$	$\neg p$	$\neg q \rightarrow \neg p$
T	T	T	F	F	T
T	F	F	T	F	F
F	T	T	F	T	T
F	F	T	T	T	T

The truth values in the columns for $p \rightarrow q$ and $\neg q \rightarrow \neg p$ are the same, thus the propositions are logically equivalent.

3. Every student at my university has a computer or has a friend who has a computer.

5. (a) False since $4 \neq -5$.

   (b) True since both $P(4)$ and $Q(4)$ are false.

   (c) True. For example, $x = 4$.

   (d) False. For example, if $x = 4$, then both $P(4)$ and $Q(4)$ are false, so the disjunction is false.

   (e) True. For example, if $x = -5$, then both $P(x)$ and $Q(x)$ are true, so the conjunction is true.

   (f) True. If $P(x)$ holds then $x = -5$, then $x^2 = 25$, so $Q(x)$ holds, so the implication holds for any x.

   (g) True. For example, if $x = -5$, then both $P(x)$ and $Q(x)$ are true, so the biconditional is true.

   (h) False. For example, if $x = 5$, then $P(x)$ is false and $Q(x)$ is true, so the biconditional is false.

7. The statement $\forall x \exists y (x + y = 0)$ is true because for any x, let $y = -x$, then $x + y = 0$. The statement $\exists y \forall x (x + y = 0)$ is false because there is no value of y such that for any x the equality $x + y = 0$ holds: for any y, consider $x = -y + 1$, then $x + y = 1$, so $x + y \neq 0$.

9. (a) True. Example: $x = 1$, $y = 2$, $1 < 2$.

   (b) True. For any x, if we take $y = x + 1$, then $x < y$.

   (c) False. There is no such x that for any y, $x < y$, because for any x we can take $y = x$, then $x \not< y$.

   (d) False. Counterexample: $x = 2$, $y = 1$, $2 \not< 1$.

   (e) False. Counterexample: $x = -1$, $1 \not< -1$.

11. (a) Not logically equivalent. Consider $P(x) = $"$x = 2$" with domain being the set of real numbers. Then $\forall x (\neg P(x))$ means $\forall x (x \neq 2)$ and is false, while $\neg (\forall x P(x))$ is true since $\forall x P(x)$, or $\forall x (x = 2)$, is false.

   (c) Logically equivalent. The proposition $\forall x (P(x) \wedge Q(x))$ is true if and only if $P(x) \wedge Q(x)$ is always true, if and only if both $P(x)$ and $Q(x)$ are always true, or $(\forall x P(x)) \wedge (\forall x Q(x))$.

   (e) Not logically equivalent. Consider $P(x) = $"$x$ is an odd integer" and $Q(x) = $"$x$ is an even integer" with domain being the set of integers. Then $\forall x (P(x) \leftrightarrow Q(x))$ is false (e.g. for $x = 1$, $P(x)$ is true and $Q(x)$ is false) while $(\forall x P(x)) \leftrightarrow (\forall x Q(x))$ is true since both $\forall x P(x)$ and $\forall x Q(x)$ are false.

(g) Logically equivalent. The proposition $\exists x(P(x) \lor Q(x))$ is true if and only if for at least one value of x, $P(x) \lor Q(x)$ is true. i.e. for at least one value of x at least one of $P(x)$ and $Q(x)$ holds, i.e. at least one of $\exists x P(x)$ and $\exists x Q(x)$ holds.

(i) Not logically equivalent. Consider $P(x) =$"x is an even integer" and $Q(x) =$"$x^2 = -1$" with domain being the set of integers. Then $\exists x(P(x) \to Q(x))$ is true (e.g. for $x = 1$, both $P(x)$ and $Q(x)$ are false, so the implication is true) while $(\exists x P(x)) \to (\exists x Q(x))$ is false (because $\exists x P(x)$ is true and $\exists x Q(x)$ is false, so the implication is false).

13. The definition is as follows: the sequence $a_1, a_2, \ldots$ converges to a number L if for any positive ε there exists an index N such that for any $n \geq N$, $|a_n - L| < \varepsilon$. We rewrite this definition using quantifiers: $\exists L \forall \varepsilon((\varepsilon > 0) \to (\exists N \forall n((n \geq N) \to (|a_n - L| < \varepsilon))))$ (where L and ε are real numbers, and n and N are natural numbers). This can also be expressed as: $\exists L \in \mathbb{R} \ \forall \varepsilon > 0 \ \exists N \in \mathbb{N} \ \forall n \geq N \ |a_n - L| < \varepsilon$.

17.3   Types of proofs

1. We will prove the statement by contrapositive: Suppose n is odd. Then $n = 2k + 1$ for some integer k. Then $3n + 5 = 3(2k + 1) + 5 = 6k + 8 = 2(3k + 4)$ is even. Thus we have proved that if n is odd, then $3n + 5$ is even. Therefore if $3n + 5$ is odd then n is even.

3. The statement is false; $n = 3$ is a counterexample, since $2^3 + 1 = 9$ is not prime.

5. By problem 2, an integer is even if and only if its square is even. Therefore an integer is odd if and only if its square is odd. Thus If an odd integer N is a perfect square, then $N = m^2$ for some odd integer m. Thus $m = 2k + 1$ for some integer k. Then $N = m^2 = (2k + 1)^2 = 4k^2 + 4k + 1 = 4(k^2 + k) + 1$, so N is of the form $4n + 1$. This proof is direct.

The converse is "if an odd number has the form $4n + 1$, then it is a perfect square". This is false because for example $5 = 4 \cdot 1 + 1$ but 5 is not a perfect square. Thus 5 is a counterexample.

7. (a) Let $f(x) = x^{101} + x^{51} + x + 1$. Then $f(-1) = -2 < 0$ and $f(1) = 4 > 0$. Since $f(x)$ is a continuous function, by the intermediate value theorem $f(x)$ has a root. This proof is nonconstructive because we did not construct a root, we only proved its existence.

(b) Suppose $f(x)$ has two distinct roots. By the mean value theorem there is a number c between these roots such that $f'(c) = 0$. But $f'(x) = 101x^{100} + 51x^{50} + 1 > 0$ for all x. We get a contradiction. This is a proof by contradiction.

9. The value $x = \dfrac{\pi}{6}$ is a root of the equation. This is a constructive proof since we provided an explicit example.

11. The roots of the equation $x^2 + x + 1 = 0$ can be found using the quadratic formula: $x = \dfrac{-1 \pm \sqrt{-3}}{2}$. Both roots are complex numbers. Since any quadratic equation has exactly two roots (counting multiplicity), there are no other roots. In particular, there are no rational (or any real) solutions. This proof is direct.

13. An integer divisible by 8 has the form $8n$ where n is an integer, and $8n = (4n^2 + 4n + 1) - (4n^2 - 4n + 1) = (2n + 1)^2 - (2n - 1)^2$. This proof is direct and constructive: we gave an explicit example of two perfect squares whose difference is equal to $8n$.

15. Proof by contrapositive: suppose that at least one of n and m is even. Without loss of generality, we can assume that n is even. Then $n = 2k$ for some integer k. Then $nm + 2n + 2m = 2km + 4k + 2m = 2(km + 2k + m)$ is even.

17.4   Principle of Mathematical Induction

1. (a) We will prove this formula by Mathematical Induction.
   Basis step: for $n = 1$ we have $1^2 = \dfrac{1 \cdot 2 \cdot 3}{6}$ which is true.
   Inductive step: suppose the formula holds for $n = k$, i.e.

   $$1^2 + 2^2 + 3^2 + \ldots + k^2 = \frac{k(k + 1)(2k + 1)}{6}.$$

   Adding $(k + 1)^2$ to both sides gives
   $$1^2 + 2^2 + 3^2 + \ldots + k^2 + (k + 1)^2 = \frac{k(k + 1)(2k + 1)}{6} + (k + 1)^2 =$$

   $$\frac{k(k + 1)(2k + 1) + 6(k + 1)^2}{6} = \frac{(k + 1)(k(2k + 1) + 6(k + 1))}{6} =$$

   $$\frac{(k + 1)(2k^2 + k + 6k + 1)}{6} = \frac{(k + 1)(2k^2 + 7k + 1)}{6} =$$

   $$\frac{(k + 1)(k + 2)(2k + 3)}{6} = \frac{(k + 1)((k + 1) + 1)(2(k + 1) + 1)}{6}.$$

   Thus the formula holds for $n = k + 1$.

   (c) Proof by Mathematical Induction.
   Basis step: for $n = 1$ we have $1 \cdot 1! = 2! - 1$, or $1 = 2 - 1$ which is true.
   Inductive step: suppose the formula holds for $n = k$, i.e.

   $$1 \cdot 1! + 2 \cdot 2! + \ldots + k \cdot k! = (k + 1)! - 1.$$

   Adding $(k + 1) \cdot (k + 1)!$ to both sides gives
   $1 \cdot 1! + 2 \cdot 2! + \ldots + k \cdot k! + (k + 1) \cdot (k + 1)! = (k + 1)! - 1 + (k + 1) \cdot (k + 1)! = (k + 1)!(1 + k + 1) - 1 = (k + 2)! - 1$.
   Thus the formula holds for $n = k + 1$.

   (e) Proof by Mathematical Induction.
   Basis step: for $n = 1$ we have $1 = 1^2$ which is true.
   Inductive step: suppose the formula holds for $n = k$, i.e.

   $$1 + 3 + 5 + \ldots + (2n - 1) = n^2.$$

   Adding $2(n + 1) - 1$ to both sides gives
   $1 + 3 + 5 + \ldots + (2n - 1) + (2(n + 1) - 1) = n^2 + (2(n + 1) - 1) = n^2 + 2n + 1 = (n + 1)^2$.
   Thus the formula holds for $n = k + 1$.

3. Proof by Mathematical Induction.
   If $q = 1$, then it is true that $3^{2^q} - 1 = 3^2 - 1 = 8$ is divisible by $2^{q+2} = 2^3 = 8$.

Assume that the statement holds for $q = k$, i.e. $3^{2^k} - 1$ is divisible by 2^{k+2}. We want to prove that the statement holds for $q = k + 1$, i.e. $3^{2^{k+1}} - 1$ is divisible by 2^{k+3}. We have: $3^{2^{k+1}} - 1 = 3^{2^k \cdot 2} - 1 = \left(3^{2^k}\right)^2 - 1 = \left(3^{2^k} - 1\right)\left(3^{2^k} + 1\right)$. By the induction hypothesis, $3^{2^k} - 1$ is divisible by 2^{k+2}. Clearly, $3^{2^k} + 1$ is an even number, thus it is divisible by 2. Then the product $\left(3^{2^k} - 1\right)\left(3^{2^k} + 1\right)$ is divisible by $2^{k+2} \cdot 2 = 2^{k+3}$.

5. (a) Proof by Mathematical Induction.
   Basis step. If $n = 1$, the identity says that $F_1 F_2 = F_2^2$, i.e. $1 \cdot 1 = 1^2$ which is true.
   Inductive step. Assume the identity holds for $n = k$, i.e.

   $$F_1 F_2 + F_2 F_3 + \ldots + F_{2k-1} F_{2k} = F_{2k}^2. \tag{17.1}$$

   We want to prove that it holds for $n = k + 1$, i.e.

   $$F_1 F_2 + F_2 F_3 + \ldots + F_{2(k+1)-1} F_{2(k+1)} = F_{2(k+1)}^2,$$

   or, equivalently,

   $$F_1 F_2 + F_2 F_3 + \ldots + F_{2k+1} F_{2k+2} = F_{2k+2}^2.$$

   Using (17.1) we have:
   $F_1 F_2 + F_2 F_3 + \ldots + F_{2k+1} F_{2k+2} = F_1 F_2 + F_2 F_3 + \ldots + F_{2k-1} F_{2k} + F_{2k} F_{2k+1} +$
   $F_{2k+1} F_{2k+2} = F_{2k}^2 + F_{2k} F_{2k+1} + F_{2k+1} F_{2k+2} = F_{2k}(F_{2k} + F_{2k+1}) + F_{2k+1} F_{2k+2} =$
   $F_{2k} F_{2k+2} + F_{2k+1} F_{2k+2} = (F_{2k} + F_{2k+1}) F_{2k+2} = F_{2k+2}^2.$
   Thus the identity holds for $n = k + 1$.

   (c) Proof by Mathematical Induction.
   Basis step. For $n = 1$ the identity is $F_0 F_2 = F_1^2 + (-1)^1$. Since $F_0 = 0$ and $F_1 = F_2 = 1$, we have $0 \cdot 2 = 1 + (-1)$ which is true.
   Inductive step. Assume the identity holds for $n = k$, i.e.

   $$F_{k-1} F_{k+1} = F_k^2 + (-1)^k.$$

   We want to show that it then holds for $n = k + 1$, i.e.

   $$F_{(k+1)-1} F_{(k+1)+1} = F_{k+1}^2 + (-1)^{k+1},$$

   or, equivalently,

   $$F_k F_{k+2} = F_{k+1}^2 + (-1)^{k+1}.$$

   We have
   $$\begin{aligned}
   F_k F_{k+2} &= F_k(F_k + F_{k+1}) \\
   &= F_k^2 + F_k F_{k+1} \\
   &= F_{k-1} F_{k+1} - (-1)^k + F_k F_{k+1} \\
   &= F_{k+1}(F_{k-1} + F_k) + (-1) \cdot (-1)^k \\
   &= F_{k+1}^2 + (-1)^{k+1}.
   \end{aligned}$$

   Thus the identity holds for $n = k + 1$.

   (d) Hint: recall that multiplication of 2×2 matrices is defined by

   $$\begin{pmatrix} a & b \\ c & d \end{pmatrix} \begin{pmatrix} e & f \\ g & h \end{pmatrix} = \begin{pmatrix} ae + bg & af + bh \\ ce + dg & cf + dh \end{pmatrix}.$$

(e) Proof by Strong Mathematical Induction.

Basis step. If $n = 1$, then the identity says that $F_0^2 + F_1^2 = F_1^2$, or $0^2 + 1^2 = 1^2$ which is true.

Inductive step. Assume that it holds for all $1 \le n \le k$. Namely, we will use that it holds for $n = k$ and $n = k - 1$, i.e.

$$F_{k-1}^2 + F_k^2 = F_{2k-1}$$

and $F_{(k-1)-1}^2 + F_{(k-1)}^2 = F_{2(k-1)-1}$, or equivalently,

$$F_{k-2}^2 + F_{k-1}^2 = F_{2k-3}.$$

We want to prove that it holds for $n = k + 1$, i.e. $F_{(k+1)-1}^2 + F_{k+1}^2 = F_{2(k+1)-1}$, or, equivalently,

$$F_k^2 + F_{k+1}^2 = F_{2k+1}.$$

It may be easier here to work from the right hand side.

$F_{2k+1} = F_{2k} + F_{2k-1} = F_{2k-1} + F_{2k-2} + F_{2k-1} = 2F_{2k-1} + F_{2k-2} = 2F_{2k-1} + F_{2k-1} - F_{2k-3} = 3F_{2k-1} - F_{2k-3} = 3(F_{k-1}^2 + F_k^2) - (F_{k-2}^2 + F_{k-1}^2) = 3F_{k-1}^2 + 3F_k^2 - F_{k-2}^2 - F_{k-1}^2 = 2F_{k-1}^2 + 3F_k^2 - F_{k-2}^2 = 2F_{k-1}^2 + 3F_k^2 - (F_k - F_{k-1})^2 = 2F_{k-1}^2 + 3F_k^2 - F_k^2 + 2F_k F_{k-1} - F_{k-1}^2 = F_{k-1}^2 + 2F_k^2 + 2F_k F_{k-1} = F_{k-1}(F_{k-1} + F_k) + F_k(F_k + F_{k-1}) + F_k^2 = F_{k-1}F_{k+1} + F_k F_{k+1} + F_k^2 = (F_{k-1} + F_k)F_{k+1} + F_k^2 = F_k^2 + F_{k+1}^2.$

Note. The idea of the above inductive step is the following: express F_{2k+1} in terms of F_i's with i odd and less than $2k+1$, e.g. in terms of F_{2k-1} and F_{2k-3}, then use the inductive hypothesis to rewrite F_{2k-1} and F_{2k-3} as sums of squares (since we assume that the formula holds for smaller indices), and then rewrite the obtained expression in terms of F_k and F_{k+1} (because the formula we want to prove involves these terms).

7. Basis step. For $n = 1$ city there is nothing to prove because there is no "any other city". (The step $n = 2$, in which case we have 2 cities and one road between them, so one city can be reached from the other, is also acceptable in this situation.)

Inductive step. Assume the statement is true for $n = k$. We want to prove that the statement is true for $n = k+1$. Suppose we are given $k+1$ cities with roads as described in the problem. Choose one city (let's call this city N) and eliminate it and all roads from and to it for a moment. We are left with k cities. By the inductive hypothesis, there is a city that can be reached from any other city either directly or via at most one other city. Let's call it A, and let's call those cities from which there are direct roads to A group B, and the rest of the cities group C. Then from every city in group C there is a road to at least one city in group B:

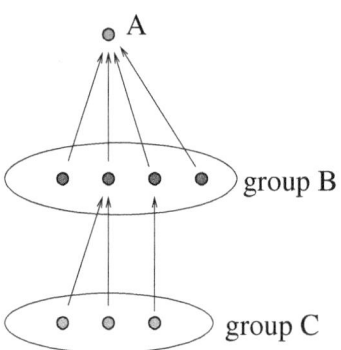

Now we add our $(k+1)$-th city N back. Consider the following 3 cases:

Case I. The road between A and N goes from N to A.

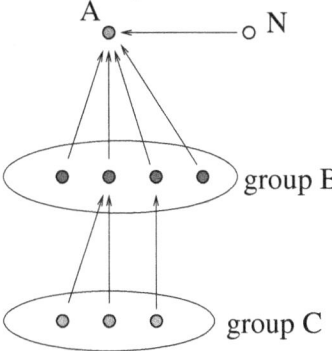

Then we put N into group B, and A is still "a solution city".

Case II. There is at least one road from N to group B.

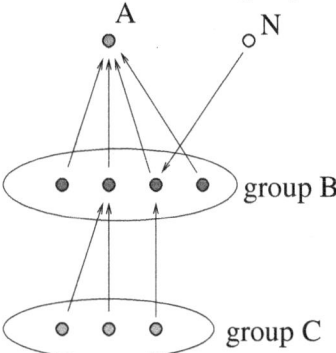

Then we put N into group C, and A is still a solution city.

Case III. None of the above: the road between A and N goes from A to N, and all the roads between group B and N lead to N.

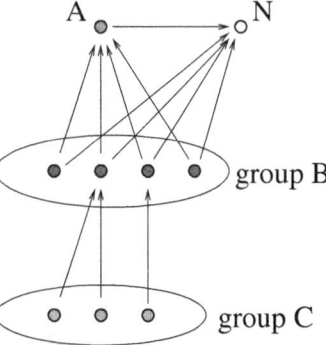

Then N is a new solution city, and A will join group B.

9. First of all, if at least one vertex has odd degree, then there is an odd number of regions around it, and it is obvious that they can not be properly colored with two colors.

   We will show that if the degree of each vertex is even, then the map can be properly colored with two colors. We will use Strong Mathematical Induction, and the induction will be on the number of line segments.

Basis step. For $n = 0$ line segments, the whole plane is one big region. We can color it with any color we like.

Inductive step. Suppose any map with less than or equal to k line segments can be properly colored with two colors. We wish to show that any map with $k+1$ line segments can be properly colored. Suppose we are given such a map. Remove temporarily all the boundary lines of any one region.

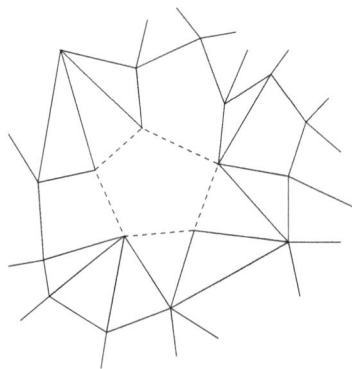

We get a map with less than k line segments, and the degree of each vertex is still even. By the inductive assumption this new map can be properly colored. Consider a coloring,

put the boundary of our region back, and change the color inside it. We get a proper coloring for our original map with $k + 1$ line segments:

Remark. It is possible that upon eliminating the boundary of a region the degree of some point(s) becomes zero. Then we have to temporarily remove these points as well.

11. First we will try to estimate the sum by estimating each term. We see that

$$\frac{1}{3n+1} \leq \text{each term} \leq \frac{1}{n+1},$$

and there are $2n+1$ terms, therefore we have

$$\frac{2n+1}{3n+1} \leq \text{sum} \leq \frac{2n+1}{n+1}.$$

The left inequality doesn't help us, but from the right one we have

$$\text{sum} \leq \frac{2n+1}{n+1} < \frac{2n+2}{n+1} = 2,$$

thus we do not need Mathematical induction for this part. To show that the sum is bigger than 1, we will use Mathematical induction.

Basis step. For $n = 1$ we have to check that $1 < \frac{1}{2} + \frac{1}{3} + \frac{1}{4}$. We calculate $\frac{1}{2} + \frac{1}{3} + \frac{1}{4} = \frac{13}{12}$, and we see that this is indeed bigger than 1.

Inductive step. Assume the inequality holds for $n = k$, i.e.

$$1 < \frac{1}{k+1} + \frac{1}{k+2} + \cdots + \frac{1}{3k+1}. \tag{17.2}$$

We want to prove that it holds for $n = k + 1$:

$$1 < \frac{1}{(k+1)+1} + \frac{1}{(k+1)+2} + \cdots + \frac{1}{3(k+1)+1},$$

or

$$1 < \frac{1}{k+2} + \frac{1}{k+3} + \cdots + \frac{1}{3k+1} + \frac{1}{3k+2} + \frac{1}{3k+3} + \frac{1}{3k+4}. \tag{17.3}$$

Compare (17.2) and (17.3), and notice that we "lost" the term $\frac{1}{k+1}$ but "gained" 3 terms: $\frac{1}{3k+2}$, $\frac{1}{3k+3}$, and $\frac{1}{3k+4}$. If we can show that we gained more than we lost, then the new sum (for $k+1$) is bigger than 1. Thus we want to show that

$$\frac{1}{3k+2} + \frac{1}{3k+3} + \frac{1}{3k+4} > \frac{1}{k+1}.$$

The following inequalities are equivalent:

$$\frac{1}{3k+2} + \frac{1}{3k+3} + \frac{1}{3k+4} > \frac{3}{3k+3}$$

$$\frac{1}{3k+2} + \frac{1}{3k+4} > \frac{2}{3k+3}$$

$$\frac{6k+6}{(3k+2)(3k+4)} > \frac{2}{3k+3}$$

$$\frac{3k+3}{(3k+2)(3k+4)} > \frac{1}{3k+3}$$

$$(3k+3)^2 > (3k+2)(3k+4)$$

$$9k^2 + 18k + 9 > 9k^2 + 18k + 8,$$

and the last one is obviously true.

13. Proof by Strong Mathematical Induction.

Basis step: for $n = 1$ we have $\det M_1 = \det [5] = 5 = \dfrac{15}{3} = \dfrac{1}{3}(4^2 - 1)$.

Inductive step: assume the formula $\det M_n = \dfrac{1}{3}(4^{n+1} - 1)$ holds for $1 \le n \le k$. We wish to prove it for $n = k + 1$.

Case I: if $k = 1$, then $k + 1 = 2$ and $\det M_2 = \det \begin{bmatrix} 5 & 2 \\ 2 & 5 \end{bmatrix} = 5^2 - 2^2 = 21 = \dfrac{63}{3} = \dfrac{1}{3}(4^3 - 1)$.

Case II: if $k \ge 2$, then $k - 1 \ge 1$, and we can assume that the formula holds for $n = k$ and $n = k - 1$. Expanding M_{k+1} across the first row and then expanding the second of the two obtained matrices down the first column gives

$$\det M_{k+1} = \det \begin{bmatrix} 5 & 2 & & & & \\ 2 & 5 & 2 & & 0 & \\ & 2 & 5 & & & \\ & & & \cdots & & \\ & 0 & & & 5 & 2 \\ & & & & 2 & 5 \end{bmatrix}_{(k+1)\times(k+1)} =$$

$$5 \det \begin{bmatrix} 5 & 2 & & & & \\ 2 & 5 & 2 & & 0 & \\ & 2 & 5 & & & \\ & & & \cdots & & \\ & 0 & & & 5 & 2 \\ & & & & 2 & 5 \end{bmatrix}_{k\times k} - 2 \det \begin{bmatrix} 2 & 2 & & & & \\ 0 & 5 & 2 & & 0 & \\ & 2 & 5 & & & \\ & & & \cdots & & \\ & 0 & & & 5 & 2 \\ & & & & 2 & 5 \end{bmatrix}_{k\times k} =$$

$$5 \det M_k - 2^2 \det \begin{bmatrix} 5 & 2 & & & & \\ 2 & 5 & 2 & & 0 & \\ & 2 & 5 & & & \\ & & & \cdots & & \\ & 0 & & & 5 & 2 \\ & & & & 2 & 5 \end{bmatrix}_{(k-1)\times(k-1)} =$$

$$5 \det M_k - 4 \det M_{k-1} = 5 \cdot \dfrac{1}{3}(4^{k+1} - 1) - 4 \cdot \dfrac{1}{3}(4^k - 1) = \dfrac{1}{3}(5 \cdot 4^{k+1} - 5 - \cdot 4^{k+1} + 4) = \dfrac{1}{3}(4 \cdot 4^{k+1} - 1) = \dfrac{1}{3}(4^{k+2} - 1).$$

15. Proof by Mathematical Induction.

Basis step. A 2×2 board with one square removed has the shape of an L-tromino, and thus can be covered by one L-tromino.

Inductive step. Assume that a $2^k \times 2^k$ board with any square removed can be covered by L-trominoes. Now suppose we are given a $2^{k+1} \times 2^{k+1}$ board with one square removed. Divide this board into four $2^k \times 2^k$ boards. One of them has one square removed, and the three others are whole boards. Temporarily remove corner squares from those three whole boards as shown on the picture below.

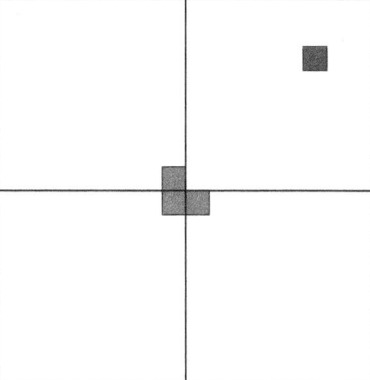

By the induction assumption, each of these four boards can be covered by L-trominoes. Now place one more L-tromino in the center to cover the 3 squares that we temporarily removed. We are done.

17.5   Dirichlet's Box Principle

1. Think of months as "boxes" and people's birth dates as "objects". Since there are more people (13) than months (12), By Dirichlet's Box Principle, at least two birth dates ("objects") are in the same month ("box").

3. There are 100 possible remainders upon division by 100 (namely, 0, 1, 2, ..., 99). Since we have 120 (more than 100) numbers, by Dirichlet's Box Principle there are at least two numbers with the same remainder. Their difference has remainder 0, and thus is divisible by 100. Since all numbers are distinct, the difference is nonzero, and therefore it ends in two zeros.

5. Divide the hexagon into 6 regions as shown in the figure below. Since we have 7 (more than 6) points, by Dirichlet's Box Principle there is a region with at least two points in it (or on its boundary). The distance between those two points is at most 1 because each region is an equilateral triangle with all sides of length 1.

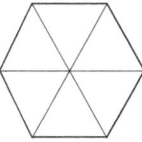

6. Hint: show that we can choose 2 of the given points whose midpoint is a lattice point.

7. There are 11 possible remainders upon division by 11 (namely, 0, 1, 2, ..., 10). Since we have 12 (more than 11) numbers, by Dirichlet's Box Principle at least two numbers have the same remainder. Their difference is then divisible by 11. Since the given numbers are two-digit, this difference is at most two-digit. Since the numbers are distinct, the difference is non-zero. There are no one-digit numbers divisible by 11. Every two-digit number that is divisible by 11, has the form aa (such numbers are 11, 22, 33, ..., 99).

9. Since $7 \cdot 7 \cdot 7 = 343$, we can divide the cube into 343 small cubes, each with edge 1. Each point is inside at most one small cube (if a point is on the boundary of a small cube, then it is not inside any small cube). Since there are more small cubes than points, there is a small cube (moreover, there are at least 43 of them) that does not contain any points inside it.

11. There are 2 possible remainders upon division by 2 (namely, 0 and 1). Since we have 3 numbers, at least two of them have the same remainder. Their difference is even, therefore the product $(x_1 - x_2)(x_1 - x_3)(x_2 - x_3)$ is even.

12. Notation clarification:
$$\prod_{1 \leq i < j \leq 4} (a_i - a_j) = (a_1 - a_2)(a_1 - a_3)(a_1 - a_4)(a_2 - a_3)(a_2 - a_4)(a_3 - a_4).$$

13. (a) Divide the rectangle into six 3×6 rectangles as shown below. Since there are 7 points, by Dirichlet's Box Principle at least two of them are in the same 3×6 rectangle. The distance between them is at most the length of the diagonal of a 3×6 rectangle, which is $\sqrt{3^2 + 6^2} = \sqrt{45} < 7$.

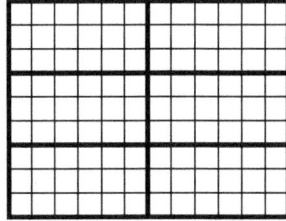

(b) Answer: yes.

Note: Although the proof in part (a) does not work for 6 points, it does not mean that the statement does not hold for 6 points.

Hint: Divide the rectangle into 5 regions with the property that the distance between any two points in one region is less than 7. This is harder than part (a), but is possible!

15. Each year contains at least 365 days. Since $365 = 52 \cdot 7 + 1$, each year contains 52 whole weeks, and thus at least 52 Fridays. Kevin is paid every other Friday, therefore he is paid at least $\dfrac{52}{2} = 26$ times a year. There are 12 months, and $26 = 2 \cdot 12 + 2$. By the Generalized Dirichlet's Box Principle, at least one month contains three days on which Kevin is paid.

17. "Make" a box for each side. There will be $2n$ boxes. We will "put" a diagonal into a box if it is parallel to the corresponding side.

box 1: diagonals parallel to side 1	. . .	box $2n$: diagonals parallel to side $2n$

We will figure out the maximal possible number of diagonals that can be parallel to one side (and thus parallel among themselves), i.e. the maximal possible number of diagonals in each box, and we will figure out how many diagonals we have in a $2n$-gon. We will show that $2n$ times the maximal number of diagonals in each box is less than the number of diagonals in a $2n$-gon, thus there is not enough space for all the diagonals in our boxes. Therefore, there is a diagonal that is not in any box, and thus not parallel to any side.

Let p be the maximal possible number of diagonals parallel to the same side. We will find a condition on p. Notice that the vertices of these p diagonals and the 2 vertices of the side they are all parallel to, are distinct (because if 2 line segments have a common vertex, they can not be parallel unless they lie on one line which is impossible in our case). Let us draw the $2n$-gon so that all these p diagonals and the parallel side are vertical, with the side on the left. Then we must also have at least one vertex on the right (because the rightmost line segment must be inside the $2n$-gon):

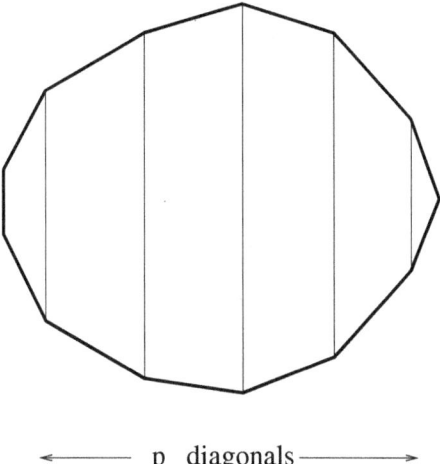

$\longleftarrow$ —— p diagonals —— $\longrightarrow$

Thus the number of vertices in this figure is at least $2p+2+1$. So we have $2n \geq 2p+3$. Since $2n$ is an even number and $2p+3$ is odd, we must have $2n \geq 2p+4$. Then $2n-4 \geq 2p$, and $n-2 \geq p$. So there may be at most $n-2$ diagonals in the same box.

Next let's count the number of diagonals in a $2n$-gon. For every diagonal, there are $2n$ ways to choose the first vertex. Once the first vertex has been chosen, there are $2n-3$ ways to choose the second vertex (because the first vertex and its immediate neighbors can not be chosen as the second vertex). Thus there are $2n(2n-3)$ ways to choose an ordered pair of vertices of a diagonal. But this way we counted each diagonal twice:

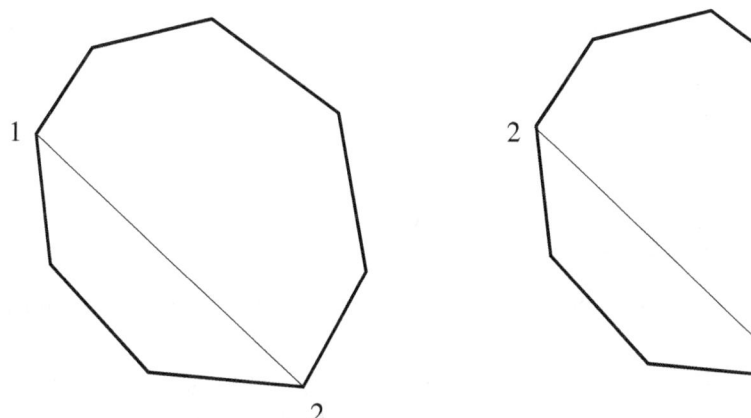

So we must divide by 2. Therefore there are $\dfrac{2n(2n-3)}{2} = n(2n-3)$ diagonals.

Thus we have $2n$ boxes, at most $n-2$ diagonals may be in the same box, therefore at most $2n(n-2) = 2n^2 - 4n$ diagonals may be in the boxes. But we have $n(2n-3) = 2n^2 - 3n$ diagonals. Since $2n^2 - 3n > 2n^2 - 4n$, there is a diagonal which is not in any box, and thus is not parallel to any side.

19. Since we have 3 rows and only 2 colors, every column has some color repeated. Since there are 7 columns, there is a color that is repeated (at least twice) in at least 4 columns. So each of these 4 columns contains at least 2 blocks of that color. If there is

a column that contains 3 blocks of that color, then choose any other of the 4 columns, and we'll have a rectangle, e.g. as shown in the picture below.

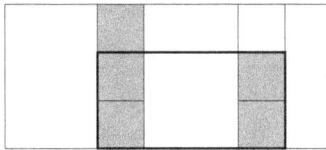

If no column contains 3 blocks of that color, then notice that there are 3 different ways to have 2 out of 3 blocks of some color:

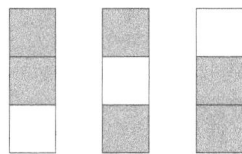

Since we have 4 columns, at least two of them have the same distribution of the colors. Then, again, we have a desired rectangle, e.g. as shown below.

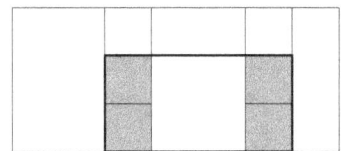

21. Consider the following numbers: $1, 11, 111, \ldots, \underbrace{11\ldots1}_{n+1}$. Since there are n possible remainders upon division by n, at least two of the above numbers, say, $\underbrace{11\ldots1}_{k}$ and $\underbrace{11\ldots1}_{l}$ for $k < l$, have the same remainder. Their difference is $\underbrace{11\ldots1}_{l} - \underbrace{11\ldots1}_{k} = \underbrace{11\ldots1}_{l-k}\underbrace{00\ldots0}_{k} = \underbrace{11\ldots1}_{l-k}\cdot10^{k}$ and is divisible by n. Since n is not divisible by 2 or 5, n and 10 are relatively prime. Therefore $\underbrace{11\ldots1}_{l-k}$ is divisible by n.

23. If $a_{i+1} - a_i < 15$ for all $1 \le i \le 7$, then $a_8 - a_1 = \sum_{i=1}^{7}(a_{i+1} - a_i) \le 7 \cdot 14 = 98$ (because all a_i's are integers, so all differences are also integers). However, this is impossible since $a_7 - a_1 = 100 - 1 = 99$.

25. Divide the numbers $\{1, 2, \ldots, 2n\}$ into n pairs of consecutive integers: $\{1, 2\}$, $\{3, 4\}$, $\ldots$, $\{2n - 1, 2n\}$. Since there are $n + 1$ numbers in the set $\{a_1, a_2, \ldots, a_{n+1}\}$, at least two of them are consecutive. Their greatest common divisor is 1 because if a natural number p divides both k and $k + 1$, then p divides their difference $(k + 1) - k = 1$, so $p = 1$. Thus our two consecutive numbers are relatively prime.

27. Since $f : X \to X$ is onto, it is a 1-1 correspondence from X onto itself, i.e. it permutes the elements of X. Then so does f^k for any integer k. If follows that f^k has an inverse f^{-k}. Since $|X| = n$, there are $n!$ permutations on X. Consider functions f, f^2, f^3, $\ldots$, $f^{n!+1}$. Since there are more function than permutations, at least two of these functions, say, f^l and f^m for $l < m$, define the same permutation, i.e. $f^l(x) = f^m(x)$ for all $x \in X$. Then $f^{m-l}(x) = f^{-l}(f^m(x)) = f^{-l}(f^l(x)) = f^{l-l}(x) = x$ for all $x \in X$.

17.6 Number theory

1. Suppose that the number $\sqrt[3]{25}$ is rational. Then it can be written as an irreducible quotient: $\sqrt[3]{25} = \dfrac{m}{n}$, $m, n \in \mathbb{Z}$, $(m, n) = 1$. Then $25 = \dfrac{m^3}{n^3}$, so $25n^3 = m^3$. Now there are several ways to get a contradiction.

   Way 1. From the last equation, $5 | m$, so $m = 5a$ for some integer a. Then $25n^3 = (5a)^3$, so $25n^3 = 125a^3$, thus $n^3 = 5a^3$. Now we see that $5 | n$. Thus both m and n are divisible by 5, which contradicts the condition $(m, n) = 1$.

   Way 2. If $n = 1$, then $25 = m^3$ which is impossible. If $n > 1$, then $n | m$ which contradicts $(m, n) = 1$.

   Way 3. We have $5 \cdot 5 \cdot n^3 = m^3$. Both n and m can be written as products of primes. Since n and m are cubed, the number of 5's on the left is 2 plus a multiple of 3, and the number of 5's on the right is a multiple of 3. This contradicts the fundamental theorem of arithmetic.

3. (a) The sum of the digits of a number

$$N = \overline{a_n a_{n-1} \ldots a_1 a_0} = a_n \cdot 10^n + a_{n-1} \cdot 10^{n-1} + \ldots + a_1 \cdot 10 + a_0 = \sum_{k=0}^{n} a_k \cdot 10^k$$

   is

$$S = a_n + a_{n-1} + \ldots + a_1 + a_0 = \sum_{k=0}^{n} a_k.$$

   Since $10 \equiv 1 \pmod 9$, $10^k \equiv 1 \pmod 9$. Then $a_k \cdot 10^k \equiv a_k \pmod 9$, and $\sum_{k=0}^{n} a_k \cdot 10^k \equiv \sum_{k=0}^{n} a_k \pmod 9$. So $N \equiv S \pmod 9$. Thus N is divisible by 9 if and only if S is divisible by 9.

   (b) If the sum of the digits of a number is 66, then the number is divisible by 3 (see problem 2(b) in chapter 1) but not divisible by 9. But if a perfect square is divisible by 3 then it must be divisible by 9. Therefore a number with the digital sum 66 cannot be a perfect square.

5. (a) First notice that if k is the last digit of m, then the last digit of m^2 is that of k^2 because $m = 10n + k$ for some n, and $m^2 = (10n + k)^2 = 100n^2 + 20nk + k^2 = (10n^2 + 2nk) \cdot 10 + k^2$. So we consider all possible last digits and compute their squares: $0^2 = 0$, $1^2 = 1$, $2^2 = 4$, $3^2 = 9$, 4^2 ends with 6, 5^2 ends with 5, 6^2 ends with 6, 7^2 ends with 9, 8^2 ends with 4, and 9^2 ends with 1. Thus the last digit of a perfect square can be 0, 1, 4, 5, 6, or 9.

   (b) Since 3 is not listed above, a number ending with 3 cannot be a perfect square.

7. No. Assume $n = a^2$ is a perfect square that ends with 65, then it is divisible by 5. Then a is divisible by 5, and therefore n is divisible by 25. Any number divisible by 25 ends with 00, 25, 50, or 75. Thus it cannot end with 65. We get a contradiction.

9. Since $2^{100} \equiv (2^2)^{50} \equiv 4^{50} \equiv (-1)^{50} \equiv 1 \pmod 5$, the remainder is 1.

   Note. There are many other ways to obtain this answer, e.g. $2^{100} \equiv (2^4)^{25} \equiv 16^{25} \equiv 1^{25} \equiv 1 \pmod 5$.

11. Since $2^4 = 16 \equiv 1 \pmod 5$ and $3^4 = 81 \equiv 1 \pmod 5$, $2^{457} + 3^{457} \equiv 2^{456+1} + 3^{456+1} \equiv$
    $2^{456} \cdot 2 + 3^{456} \cdot 3 = 2^{4 \cdot 114} \cdot 2 + 3^{4 \cdot 114} \cdot 3 \equiv (2^4)^{114} \cdot 2 + (3^4)^{114} \cdot 3 \equiv 1 \cdot 2 + 1 \cdot 3 \equiv 0 \pmod 5$.

13. If the units digit of n is 3 then n can be written in the form $n = 10k + 3$ for some integer
    k. Then $n^2 + 1 = (10k + 3)^2 + 1 = 100k^2 + 60k + 9 + 1 = 100k^2 + 60k + 10 = 5(20k^2 + 12k + 2)$
    is divisible by 5.

15. If n is composite, then $n = ab$ for some $1 < a, b < n$. Then

    $$2^n - 1 = (2^a)^b - 1^b = (2^a - 1)((2^a)^{b-1} + \ldots + 2^a + 1).$$

    Both factors are bigger than 1: $2^a - 1 > 2^1 - 1 = 1$, and $(2^a)^{b-1} + \ldots + 2^a + 1 > 1$, so
    $2^n - 1$ is composite.

17. Solution 1. Rewrite the equation as $xy - x - y = 0$. Adding 1 to both sides gives
    $xy - x - y + 1 = 1$, and factoring the left hand side gives $(x - 1)(y - 1) = 1$. Since
    $x - 1$ and $y - 1$ are integers and their product is equal to 1, they are either both 1 or
    both -1. If they are both 1, then $x = 2$ and $y = 2$. If they are both -1, then $x = 0$
    and $y = 0$. Thus we have two solution pairs.

    Solution 2. Rewrite the equation as $y(x - 1) = x$. Consider the following two cases.
    Case I: $x - 1 = 0$, then $x = 1$. The equation becomes $0 = 1$ which is impossible.
    Case II: $x - 1 \neq 0$. Then $y = \dfrac{x}{x - 1}$. If y is an integer, then $(x-1)|x$. Since $(x-1)|(x-1)$,
    $(x - 1)|(x - (x - 1))$, i.e. $(x - 1)|1$. Then either $x - 1 = 1$ or $x - 1 = -1$. Therefore
    either $x = 2$ (and then $y = 2$) or $x = 0$ (and then $y = 0$).

19. Suppose the equation has an integral solution. Consider the equation modulo 3. The
    number x is congruent to 0, 1, or 2 modulo 3. Then x^2 is congruent to either 0 or
    1 modulo 3 (because $0^2 = 0$, $1^2 = 1$, and $2^2 = 4 \equiv 1 \pmod 3$). The number $3y^2$ is
    congruent to 0 modulo 3 since $3y^2$ is divisible by 3. Thus the left hand side is congruent
    to either 0 or 1 modulo 3. But the right hand side is 17 which is congruent to 2 modulo
    3. We get a contradiction.

21. Notice first of all that y must be even because $3y = 100 - 2x$ and the right hand side
    is even. Second, y must be positive. Third, y cannot exceed 32 because if $y \geq 34$
    then $3y \geq 102$ and then x would have to be negative. But if y satisfies all the above
    conditions, namely, y is even and $2 \leq y \leq 32$, then $3y$ is even and $6 \leq 3y \leq 96$, so
    $100 - 3y$ is even and $4 \leq 100 - 3y \leq 94$, so there exists a positive integer x such that
    $2x = 100 - 3y$. Thus for any even y such that $2 \leq y \leq 32$ we have a unique solution pair
    (x, y). There are 16 even numbers satisfying $2 \leq y \leq 32$, thus 16 pairs are solutions to
    the given equation.

23. Yes. Let's find the remainder of 1239999 upon division by 4567 and subtract the
    remainder from 1239999. We'll get a number that satisfies the required conditions.
    Since $1239999 = 4567 \cdot 271 + 2342$, the number $1237657 = 1239999 - 2342 = 4567 \cdot 271$
    is divisible by 4567.

    Note. This, of course, is not the only such number. Another one is $1237657 - 4567 = 1233090$. Also, we could have started with e.g. 1239999, etc.

17.7   Case analysis

1. *Solution 1.* Since n can be congruent to 0, 1, 2, 3, or 4 modulo 5, we have five cases:
   If $n \equiv 0 \pmod 5$, then $n^2 + 2 \equiv 2 \pmod 5$;

if $n \equiv 1 \pmod 5$, then $n^2 + 2 \equiv 3 \pmod 5$;
if $n \equiv 2 \pmod 5$, then $n^2 + 2 \equiv 6 \equiv 1 \pmod 5$;
if $n \equiv 3 \pmod 5$, then $n^2 + 2 \equiv 11 \equiv 1 \pmod 5$;
if $n \equiv 4 \pmod 5$, then $n^2 + 2 \equiv 18 \equiv 3 \pmod 5$;
We see that $n^2 + 2$ is never congruent to 0 modulo 5, so it is never divisible by 5.

Solution 2. By problem 5 in chapter 6, a perfect square n^2 can end with 0, 1, 4, 5, 6, or 9. Then $n^2 + 2$ can end with 2, 3, 6, 7, 8, or 1. Since it never ends with 0 or 5, it is never divisible by 5.

3.   (a) First check $x = 0$. This is not a root because 0^0 is undefined. If $x \neq 0$, we can divide both sides of the equation by x^2. We get $x^{x^2-2} = 1$. Consider the following three cases.

      Case I: $x = 1$ is a root since (as is easy to check) it satisfies the original equation.

      Case II: $x \neq 0$, $x^2 - 2 = 0$ gives $x = \pm\sqrt{2}$. Neither of these is an integer.

      Case III: $x = -1$, $x^2 - 2$ is even. This has no solutions because if $x = -1$, then $x^2 - 2 = -1$ which is not even.

      Answer: 1.

  (c) Rewrite the equation as $x^{(x^x)} = x^{(x^2)}$. Check $x = 0$. This is not a root since 0^0 is undefined. If $x \neq 0$, we can divide both sides of the equation by $x^{(x^2)}$. We get $x^{x^x - x^2} = 1$. Consider the following three cases.

      Case I: $x = 1$ is a root.

      Case II: $x \neq 0$, $x^x - x^2 = 0$. This equation is equivalent to $x^x = x^2$. Since $x \neq 0$, we can divide both sides of the equation by x^2 which gives $x^{x-2} = 1$. Next we consider three cases for the equation $x^{x-2} = 1$.

      Case IIA: $x = 1$ is a root.

      Case IIB: $x \neq 0$, $x - 2 = 0$ gives $x = 2$ which is a root.

      Case IIC: $x = -1$, $x - 2$ is even. However, $x - 2$ is odd when $x = -1$, so this is not a root.

      Now let's continue considering cases for the original equation.

      Case III: $x = -1$, $x^x - x^2$ is even. Since $x^x - x^2$ is indeed even when $x = -1$, it is a root of the original equation.

      Answer: $-1, 1, 2$.

5. The given equations imply that $y^{(x+y)^2} = \left(y^{x+y}\right)^{x+y} = x^{x+y} = y^4$, so we have $y^{(x+y)^2} = y^4$. If $y = 0$, then the first equation in the system becomes $x^x = 0$ which has no solutions. If $y \neq 0$, we can divide both sides of the equation $y^{(x+y)^2} = y^4$ by y^4. We get $y^{(x+y)^2 - 4} = 1$. Consider the following three cases.

Case I: $y = 1$. Then the second equation in the system becomes $1^{x+1} = x$, so $x = 1$. It is easy to check that the pair $(1, 1)$ satisfies both equation in the system.

Case II: $y \neq 0$, $(x + y)^2 - 4 = 0$. This equation implies $(x + y)^2 = 4$, therefore either $x + y = 2$ or $x + y = -2$. Consider these two cases.

Case IIA: $x + y = 2$. Then $y = 2 - x$, and the second equation in the system becomes $(2 - x)^2 = x$, or $x^2 - 5x + 4 = 0$, or $(x - 1)(x - 4) = 0$. This equation has two roots: $x = 1$ and $x = 4$. If $x = 1$, then $y = 2 - 1 = 1$, and we get the pair $(1, 1)$ again. If $x = 4$, then $y = 2 - 4 = -2$. It is easy to check that the pair $(4, -2)$ satisfies both equations in the system.

Case IIB: $x+2 = -2$. Then $y = -2-x$, and the second equation in the system becomes $(-2-x)^2 = x$, or $x^2 + 3x + 4 = 0$. This equation has no real solutions.

Case III: $y = -1$, $(x+y)^2 - 4$ is even. If $y = -1$, the second equation in the system becomes $(-1)^{x+y} = x$. This implies that either $x = 1$ or $x = -1$. As is easy to check, the pair $(1, -1)$ satisfies both equations in the system, but the pair $(-1, -1)$ does not.

Answer: $(1, -1)$, $(1, 1)$, $(4, -2)$.

7.   (a) *Solution 1.* Consider two cases for each of the expressions inside an absolute value: $2x + 3$ and x. Thus we have four cases total.

Case I. $2x + 3 \geq 0$, $x \geq 0$. Then $|2x + 3| = 2x + 3$ and $|x| = x$, so the equation becomes $2x + 3 - x = 3$, or $x = 0$. This root satisfies both of the above conditions.

Case II. $2x + 3 \geq 0$, $x < 0$. Then $|2x + 3| = 2x + 3$ and $|x| = -x$, so the equation becomes $2x + 3 + x = 3$, or $3x = 0$. Then $x = 0$. This root does not satisfy the condition $x < 0$. (So we should disregard it here, however, it is a solution to the equation according to Case I.)

Case III. $2x + 3 < 0$, $x \geq 0$. Then $|2x + 3| = -(2x + 3)$ and $|x| = x$, so the equation becomes $-(2x + 3) - x = 3$, or $-3x = 6$. Then $x = -2$. This root does not satisfy the condition $x \geq 0$, so we disregard it.

Note. Actually, we could notice that if $x \geq 0$, then $2x + 3$ cannot be negative, so no real number would satisfy both of these conditions. So we could disregard this case from the very beginning.)

Case IV. $2x + 3 < 0$, $x < 0$. Then $|2x + 3| = -(2x + 3)$ and $|x| = -x$, so the equation becomes $-(2x + 3) + x = 3$, or $-x = 6$. Then $x = -6$. This root satisfies both of the above conditions.

Answer: -6, 0.

Solution 2. Since $2x + 3$ changes sign at $x = -\dfrac{3}{2}$ and x changes sign at $x = 0$, we will consider the following three intervals: $\left(-\infty, -\dfrac{3}{2}\right)$, $\left[-\dfrac{3}{2}, 0\right)$, and $[0, +\infty)$.

If $x \in \left(-\infty, -\dfrac{3}{2}\right)$, then $2x + 3 < 0$ and $x < 0$. Therefore $|2x + 3| = -(2x + 3)$ and $|x| = -x$. The equation becomes $-(2x + 3) + x = 3$, or $-x = 6$. Then $x = -6$. This root lies in the interval $\left(-\infty, -\dfrac{3}{2}\right)$, therefore is a solution.

If $x \in \left[-\dfrac{3}{2}, 0\right)$, then $2x + 3 \geq 0$ and $x < 0$. Therefore $|2x + 3| = 2x + 3$ and $|x| = -x$. The equation becomes $2x + 3 + x = 3$, or $3x = 0$. Then $x = 0$. However, this root does not lie in the interval $\left[-\dfrac{3}{2}, 0\right)$, so we disregard it.

If $x \in [0, +\infty)$, then $2x + 3 \geq 0$ and $x \geq 0$. Therefore $|2x + 3| = 2x + 3$ and $|x| = x$. The equation becomes $2x + 3 - x = 3$, or $x = 0$. This root lies in the interval $[0, +\infty)$, therefore is a solution.

Note. The interval $\left(-\infty, -\dfrac{3}{2}\right)$ corresponds to Case IV in Solution 1. The interval $\left[-\dfrac{3}{2}, 0\right)$ corresponds to Case II, and the interval $[0, +\infty)$ corresponds to Case I. (As was noted in Solution 1, Case III is impossible.)

Remark. The technique used in Solution 2 is especially better when there are 3 or more absolute values. Say, if three absolute values of linear functions are present, then using the technique of Solution 1 we have to consider eight cases, but using the technique of Solution 2 we have to consider only four intervals.

(c) Since $3x + 6$ changes sign at $x = -2$ and $x - 1$ changes sign at $x = 1$, we will consider the following three intervals: $(-\infty, -2)$, $[-2, 1)$, and $[1, +\infty)$.

If $x \in (-\infty, -2)$, then $3x + 6 < 0$ and $x - 1 < 0$. Therefore $|3x+6| = -(3x+6)$ and $|x - 1| = -(x - 1)$. The equation becomes $-(3x+6) - (x-1) = 2$, or $-4x - 5 = 2$. Then $x = -\dfrac{7}{4}$. This root does not lie in the interval $(-\infty, -2)$, so we disregard it.

If $x \in [-2, 1)$, then $3x + 6 \geq 0$ and $x - 1 < 0$. Therefore $|3x + 6| = 3x + 6$ and $|x - 1| = -(x - 1)$. The equation becomes $(3x + 6) - (x - 1) = 2$, or $2x + 7 = 2$. Then $x = -\dfrac{5}{2}$. This root does not lie in the interval $[-2, 1)$, so we disregard it.

If $x \in [1, +\infty)$, then $3x + 6 \geq 0$ and $x - 1 \geq 0$. Therefore $|3x + 6| = 3x + 6$ and $|x - 1| = x - 1$. The equation becomes $(3x + 6) + (x - 1) = 2$, or $4x + 5 = 2$. Then $x = -\dfrac{3}{4}$. This root does not lie in the interval $[1, +\infty)$, so we disregard it.

Answer: the equation has no solutions.

9. (a) Since $x - 5$ changes sign at $x = 5$ and $2x - 4$ changes sign at $x = 2$, we will consider the following three intervals: $(-\infty, 2)$, $[2, 5)$, and $[5, +\infty)$.

If $x \in (-\infty, 2)$, then $x - 5 < 0$ and $2x - 4 < 0$. Therefore $|x - 5| = -(x - 5)$ and $|2x - 4| = -(2x - 4)$. The inequality becomes $-(x - 5) - (2x - 4) \leq 6$, or $-3x + 9 \leq 6$. This is equivalent to $x \geq 1$. The solution set in this case is $[1, 2)$.

If $x \in [2, 5)$, then $x - 5 < 0$ and $2x - 4 \geq 0$. Therefore $|x - 5| = -(x - 5)$ and $|2x - 4| = 2x - 4$. The inequality becomes $-(x - 5) + (2x - 4) \leq 6$, or $x + 1 \leq 6$. This is equivalent to $x \leq 5$. Since each point in the interval $[2, 5)$ satisfies the condition $x \leq 5$, the solution set in this case is $[2, 5)$.

If $x \in [5, +\infty)$, then $x - 5 \geq 0$ and $2x - 4 \geq 0$. Therefore $|x - 5| = x - 5$ and $|2x - 4| = 2x - 4$. The inequality becomes $(x - 5) + (2x - 4) \leq 6$, or $3x - 9 \leq 6$. This is equivalent to $x \leq 5$. Since the only value in the interval $[5, +\infty)$ that satisfies the condition $x \leq 5$ is 5, this is the only solution in this case.

Finally, we take the union of the above solution sets.

Answer: $[1, 5]$.

(c) *Solution 1.* Since $x - 1$ changes sign at $x = 1$ and $x - 3$ changes sign at $x = 3$, we will consider the following three intervals: $(-\infty, 1)$, $[1, 3)$, and $[3, +\infty)$.

If $x \in (-\infty, 1)$, then $x - 1 < 0$ and $x - 3 < 0$. Therefore $|x - 1| = -(x - 1)$ and $|x - 3| = -(x - 3)$. The inequality becomes $-(x - 1) + (x - 3) > 5$, or $-2 > 5$. Since this inequality does not hold for any value of x, the solution set in this case is empty.

If $x \in [1, 3)$, then $x - 1 \geq 0$ and $x - 3 < 0$. Therefore $|x - 1| = x - 1$ and $|x - 3| = -(x - 3)$. The inequality becomes $(x - 1) + (x - 3) > 5$, or $2x - 4 > 5$. This is equivalent to $x > \dfrac{9}{2}$. Since this inequality does not hold for any values of x in the interval $[1, 3)$, the solution set in this case is empty.

If $x \in [3, +\infty)$, then $x - 1 \geq 0$ and $x - 3 \geq 0$. Therefore $|x - 1| = x - 1$ and $|x - 3| = x - 3$. The inequality becomes $(x - 1) - (x - 3) > 5$, or $2 > 5$. Since this inequality does not hold for any value of x, the solution set in this case is empty.

Answer: the inequality does not have any solutions.

Solution 2. Observe that $|x-1|$ is the distance between x and 1 on the real number line, $|x-3|$ is the distance between x and 3, and the left hand side $|x-1|-|x-3|$ represents the difference of these distances. It is clear that this difference is never larger than 5: if x lies to the left of 1, then the difference is -2; if x lies between 1 and 3 (inclusive), the difference is between -2 and 2 (inclusive), and if x lies to the right of 3, then the difference is 2. Therefore the inequality has no solutions.

11. (a) First let's sketch the graph of $h(x) = x + |x+2|$.

Case I: $x+2 \geq 0$, or $x \geq -2$. Then $|x+2| = x+2$, so $h(x) = x+(x+2) = 2x+2$.

Case II: $x+2 < 0$, or $x < -2$. Then $|x+2| = -(x+2)$, so $h(x) = x-(x+2) = -2$.

Thus $h(x) = \begin{cases} 2x+2 & \text{if } x \geq -2 \\ -2 & \text{if } x < -2 \end{cases}$. Its graph is shown below.

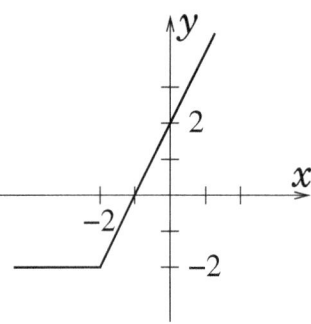

Since $f(x) = |x + |x+2|| = |h(x)|$, the graph of $f(x)$ is obtained from the graph of $h(x)$ by reflecting the piece below the x-axis about the x-axis:

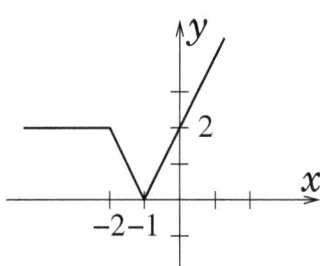

13. (a) Case I: $x \geq 0$, $y \geq 0$. The inequality becomes $x + y^3 < 8$, or $x < 8 - y^3$.

Case II: $x \geq 0$, $y < 0$. The inequality becomes $x - y^3 < 8$, or $x < 8 + y^3$.

Case III: $x < 0$, $y \geq 0$. The inequality becomes $-x + y^3 < 8$, or $x > y^3 - 8$.

Case IV: $x < 0$, $y < 0$. The inequality becomes $-x - y^3 < 8$, or $x > -8 - y^3$.

Now we draw the corresponding region in each quadrant, and we get the following figure:

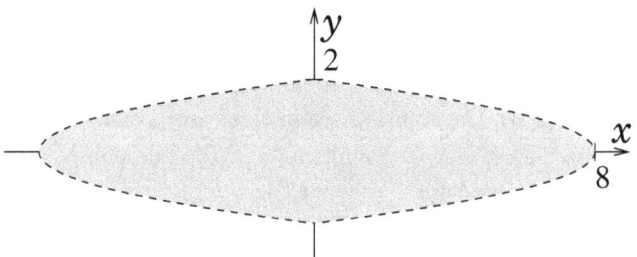

Note: since the inequality is strict, the boundary of the region is excluded.

(c) Since $y - x$ can be nonnegative or negative and $x + y$ can be nonnegative or negative, we have four cases:

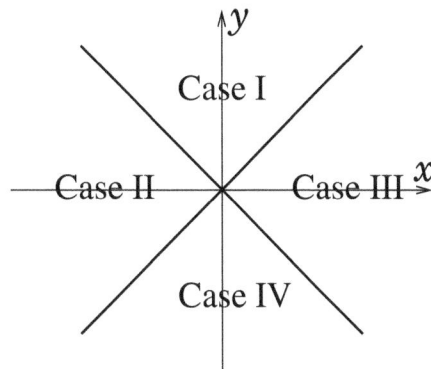

Case I: $y - x \geq 0$, $x + y \geq 0$. The inequality becomes $2y - 2x + y + x \leq 1$, or $3y \leq 1 + x$, or $y \leq \frac{1}{3} + \frac{1}{3}x$.

Case II: $y - x \geq 0$, $x + y < 0$. The inequality becomes $2y - 2x - y - x \leq 1$, or $y \leq 1 + 3x$.

Case III: $y - x < 0$, $x + y \geq 0$. The inequality becomes $-2y + 2x + y + x \leq 1$, or $-y \leq 1 - 3x$, or $y \geq 3x - 1$.

Case IV: $y - x < 0$, $x + y < 0$. The inequality becomes $-2y + 2x - y - x \leq 1$, or $-3y \leq 1 - x$, or $y \geq -\frac{1}{3} + \frac{1}{3}x$.

Now sketch the region in each case.

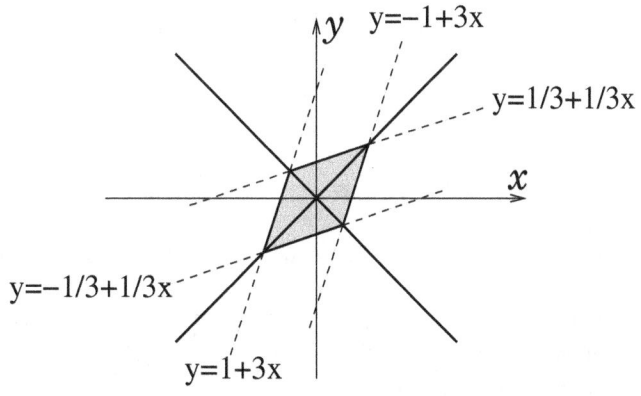

17.8   Finding a pattern

1. (a) $a_n = n^2$

   (c) $a_n = 2n + 3$

   (e) $a_n = 5 - 2n$

   (g) $a_n = \left[\dfrac{3n}{2}\right] = \begin{cases} \dfrac{3n-1}{2} & \text{if } n \text{ is odd} \\[2mm] \dfrac{3n}{2} & \text{if } n \text{ is even} \end{cases}$

(i) $a_n = \dfrac{n}{2^n}$

3. Let $S_n = \dfrac{1}{1 \cdot 3} + \dfrac{1}{3 \cdot 5} + \ldots + \dfrac{1}{(2n-1)(2n+1)}$. Then

$$S_1 = \frac{1}{1 \cdot 3} = \frac{1}{3},$$

$$S_2 = \frac{1}{1 \cdot 3} + \frac{1}{3 \cdot 5} = \frac{6}{15} = \frac{2}{5},$$

$$S_3 = \frac{2}{5} + \frac{1}{5 \cdot 7} = \frac{15}{35} = \frac{3}{7},$$

$$S_4 = \frac{3}{7} + \frac{1}{7 \cdot 9} = \frac{28}{63} = \frac{4}{9}.$$

We guess from the above calculations that $S_n = \dfrac{n}{2n+1}$. We will prove this formula by Mathematical Induction.

Basis step: if $n = 1$, the formula $S_1 = \dfrac{1}{2 \cdot 1 + 1}$ is correct.

Inductive step: suppose the formula $S_n = \dfrac{n}{2n+1}$ holds for some $n = k$, i.e. $S_k = \dfrac{k}{2k+1}$. We want to prove that it holds for $n = k+1$, i.e. $S_{k+1} = \dfrac{k+1}{2(k+1)+1}$.

Indeed, $S_{k+1} = \dfrac{1}{1 \cdot 3} + \dfrac{1}{3 \cdot 5} + \ldots + \dfrac{1}{(2k-1)(2k+1)} + \dfrac{1}{(2(k+1)-1)(2(k+1)+1)} =$

$S_k + \dfrac{1}{(2k+1)(2k+3)} = \dfrac{k}{2k+1} + \dfrac{1}{(2k+1)(2k+3)} = \dfrac{k(2k+3)+1}{(2k+1)(2k+3)} =$

$\dfrac{2k^2+3k+1}{(2k+1)(2k+3)} = \dfrac{(2k+1)(k+1)}{(2k+1)(2k+3)} = \dfrac{k+1}{2k+3} = \dfrac{k+1}{2(k+1)+1}.$

5. First we compute $f_n(x)$ for the first few values of n:
$f_1(x) = 2x + 1$,
$f_2(x) = 2(2x+1) + 1 = 4x + 3$,
$f_3(x) = 2(4x+3) + 1 = 8x + 7$,
$f_4(x) = 2(8x+7) + 1 = 16x + 15$.
It appears that $f_n(x) = 2^n x + 2^n - 1$, so we will try to prove this formula by Mathematical Induction.
Basis step: if $n = 1$, then our formula gives $f_1(x) = 2x + 1$ which is true.
Inductive step: suppose the formula holds for $n = k$, i.e. $f_k(x) = 2^k x + 2^k - 1$. Then
$f_{k+1}(x) = f_1 \circ f_k(x) = 2(2^k x + 2^k - 1) + 1 = 2^{k+1} x + 2^{k+1} - 2 + 1 = 2^{k+1} x + 2^{k+1} - 1$.
Thus the formula holds for $n = k+1$.

7. (a) As discussed in chapter 6, the units digit of a positive number is its remainder upon division by 10.

    Solution 1. First we find the units digit of 107^n for some small values of n:
$107^1 \equiv 7 \pmod{10}$,
$107^2 = 107 \cdot 107 \equiv 7 \cdot 7 \equiv 49 \equiv 9 \pmod{10}$,
$107^3 = 107^2 \cdot 107 \equiv 9 \cdot 7 \equiv 63 \equiv 3 \pmod{10}$,
$107^4 = 107^3 \cdot 107 \equiv 3 \cdot 7 \equiv 21 \equiv 1 \pmod{10}$,
$107^5 = 107^4 \cdot 107 \equiv 1 \cdot 7 \equiv 7 \pmod{10}$,
$107^6 = 107^5 \cdot 107 \equiv 7 \cdot 7 \equiv 49 \equiv 9 \pmod{10}$.

We see that the units digits start repeating. As we keep multiplying our number by 107, the 4-tuple of units digits 7, 9, 3, and 1 keep repeating. More precisely,

$$107^n \equiv \begin{cases} 7 \ (\text{mod } 10) & \text{if } n \equiv 1 \ (\text{mod } 4) \\ 9 \ (\text{mod } 10) & \text{if } n \equiv 2 \ (\text{mod } 4) \\ 3 \ (\text{mod } 10) & \text{if } n \equiv 3 \ (\text{mod } 4) \\ 1 \ (\text{mod } 10) & \text{if } n \equiv 0 \ (\text{mod } 4) \end{cases}.$$

This formula can be proved by Strong Mathematical Induction.

Basis step: if $n = 1$, then $107^n \equiv 7 \ (\text{mod } 10)$ is true.

Inductive step: Suppose the formula holds for all $1 \le n \le k$. We want to prove that it holds for $n = k + 1$.

Case I: $k + 1 = 2$. Then $k + 1 \equiv 2 \ (\text{mod } 4)$, and $107^2 \equiv 9 \ (\text{mod } 10)$ is true.

Case II: $k + 1 = 3$. Then $k + 1 \equiv 3 \ (\text{mod } 4)$, and $107^3 \equiv 3 \ (\text{mod } 10)$ is true.

Case III: $k + 1 = 4$. Then $k + 1 \equiv 0 \ (\text{mod } 4)$, and $107^4 \equiv 1 \ (\text{mod } 10)$ is true.

Case IV: $k + 1 \ge 5$. Then $k - 3 = (k + 1) - 4 \ge 1$, and we assumed that the formula above was true for $n = k - 3$.

We consider all possible remainders of $k + 1$ modulo 4.

Case IVA: $k+1 \equiv 1 \ (\text{mod } 4)$. Then $k-3 \equiv 1 \ (\text{mod } 4)$, so $107^{k+1} \equiv 107^{k-3} \cdot 107^4 \equiv 107^{k-3} \cdot 1 \equiv 7 \ (\text{mod } 10)$.

Case IVB: $k+1 \equiv 2 \ (\text{mod } 4)$. Then $k-3 \equiv 2 \ (\text{mod } 4)$, so $107^{k+1} \equiv 107^{k-3} \cdot 107^4 \equiv 107^{k-3} \cdot 1 \equiv 9 \ (\text{mod } 10)$.

Case IVC: $k+1 \equiv 3 \ (\text{mod } 4)$. Then $k-3 \equiv 3 \ (\text{mod } 4)$, so $107^{k+1} \equiv 107^{k-3} \cdot 107^4 \equiv 107^{k-3} \cdot 1 \equiv 3 \ (\text{mod } 10)$.

Case IVD: $k+1 \equiv 0 \ (\text{mod } 4)$. Then $k-3 \equiv 0 \ (\text{mod } 4)$, so $107^{k+1} \equiv 107^{k-3} \cdot 107^4 \equiv 107^{k-3} \cdot 1 \equiv 1 \ (\text{mod } 10)$.

Thus the formula holds for $n = k + 1$.

Now, since $107 \equiv 3 \ (\text{mod } 4)$, $107^{107} \equiv 3 \ (\text{mod } 10)$, so the units digit of 107^{107} is 3.

Solution 2. Since $107^4 \equiv 1 \ (\text{mod } 10)$ and $107^3 \equiv 3 \ (\text{mod } 10)$ (as we saw above), $107^{107} \equiv 107^{104} \cdot 107^3 \equiv (107^4)^{26} \cdot 107^3 \equiv 1^{26} \cdot 107^3 \equiv 1 \cdot 3 \equiv 3 \ (\text{mod } 10)$.

9. First we find the remainder of 2^n upon division by 12 for some small n: $2^1 = 2 \equiv 2 \ (\text{mod } 12)$,

$2^2 = 4 \equiv 4 \ (\text{mod } 12)$,

$2^3 = 8 \equiv 8 \ (\text{mod } 12)$,

$2^4 = 16 \equiv 4 \ (\text{mod } 12)$,

$2^5 = 32 \equiv 8 \ (\text{mod } 12)$.

We see that the remainders 4 and 8 start repeating. Namely,

$$2^n \equiv \begin{cases} 4 \ (\text{mod } 12) & \text{if } n \text{ is even} \\ 8 \ (\text{mod } 12) & \text{if } n \ge 3 \text{ is odd} \end{cases}.$$

As in problem 7, this can be proved by Strong Mathematical Induction. Since 100 is even, the remainder of 2^{100} upon division by 12 is 4.

11. (a) For $f(x) = \sin(x)$, the first few derivatives are:

$f'(x) = \cos(x)$,

$f''(x) = -\sin(x)$,

$f'''(x) = -\cos(x)$,

$f^{(4)}(x) = \sin(x)$,

$f^{(5)}(x) = \cos(x)$.

We got $\cos(x)$ again, so the derivatives $\cos(x)$, $-\sin(x)$, $-\cos(x)$, $\sin(x)$ will repeat. Therefore

$$f^{(n)}(x) = \begin{cases} \cos(x) & \text{if } n \equiv 1 \pmod 4 \\ -\sin(x) & \text{if } n \equiv 2 \pmod 4 \\ -\cos(x) & \text{if } n \equiv 3 \pmod 4 \\ \sin(x) & \text{if } n \equiv 0 \pmod 4 \end{cases}.$$

This formula can be proved by Strong Mathematical Induction (the proof is similar to that in problem 7).

(c) For $h(x) = 2e^{5x}$, the first few derivatives are:
$h'(x) = 2 \cdot 5e^{5x}$,
$h''(x) = 2 \cdot 5 \cdot 5e^{5x}$,
$h'''(x) = 2 \cdot 5 \cdot 5 \cdot 5e^{5x}$.
We guess that $h^{(n)}(x) = 2 \cdot 5^n e^{5x}$, and prove this formula by Mathematical Induction.
Basis step: $h'(x) = 2 \cdot 5e^{5x}$ is true.
Inductive step: suppose $h^{(k)}(x) = 2 \cdot 5^k e^{5x}$, then $h^{(k+1)}(x) = \left(2 \cdot 5^k e^{5x}\right)' = 2 \cdot 5^k \cdot 5e^{5x} = 2 \cdot 5^{k+1} e^{5x}$.

13. First we find the number of regions for a few small values n:

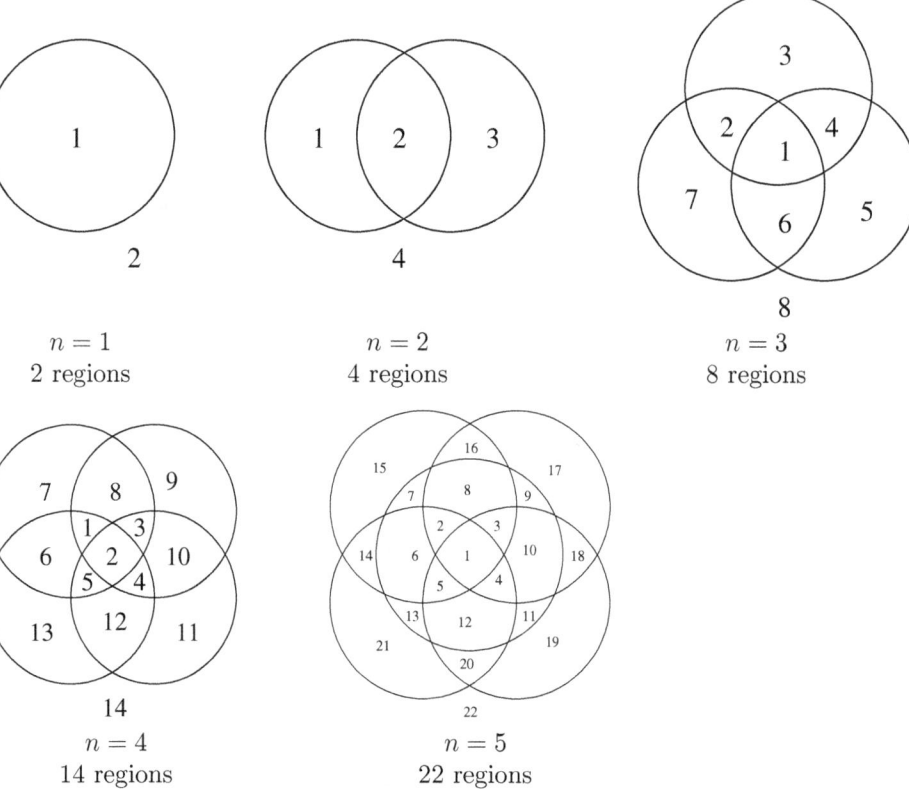

$n = 1$
2 regions

$n = 2$
4 regions

$n = 3$
8 regions

$n = 4$
14 regions

$n = 5$
22 regions

The sequence is 2, 4, 8, 14, 22, ... The differences between consecutive terms are 2, 4, 6, 8, ... We guess that the differences are increasing consecutive even numbers, so the number of regions into which n circles divide the plane is $2 + 2 + 4 + 6 + ... + 2(n-1) = 2 + 2(1 + 2 + 3 + ... + (n-1)) = 2 + 2\dfrac{(n-1)n}{2} = 2 + (n-1)n = n^2 - n + 2$.

Now we will prove this formula by Mathematical Induction.

Basis step: if $n = 1$, the formula gives 2, and it is true that there are 2 regions.

Inductive step: suppose the formula is true for $n = k$ circles. Given $k + 1$ circles, temporarily remove one circle. The remaining k circles divide the plane into $k^2 - k + 2$ regions. Now add the $(k+1)$-th circle back. This new circle intersects the old k circles in $2k$ points. Thus the intersection points divide the new circle into $2k$ arcs. Therefore, the number of regions increases by $2k$ (each arc divides an old region into 2). Then, since k circles divided the plane into $k^2 - k + 2$ regions, $k + 1$ circles will divide it into $k^2 - k + 2 + 2k = k^2 + 2k + 1 - k - 1 + 2 = (k + 1)^2 - (k + 1) + 2$ regions, and thus the formula holds for $k + 1$.

14. Hint: do not try to list all the ways. There are too many of them! Better replace 10 by small numbers, and guess the pattern. Prove your guess using Mathematical Induction.

15. We compute the first few Fibonacci numbers: 0, 1, 1, 2, 3, 5, 8, 13, 21, 34, and notice that every third Fibonacci number is even. More precisely, F_n is even if and only if $n \equiv 0 \pmod 3$. Therefore exactly one third of F_1, F_2, ..., F_{99} are even which gives 33 numbers, and F_0 is even, thus we have 34 even numbers total.

The pattern described above can be proved by Strong Mathematical Induction as follows.

Basis step: if $n = 0$, $F_0 = 0$ is even.

Inductive step: suppose the statement "F_n is even if and only if $n \equiv 0 \pmod 3$" holds for $0 \le n \le k$. We will prove that the statement holds for $n = k + 1$.

Case I: $k + 1 = 1$. Then $k + 1 \not\equiv 0 \pmod 3$, and $F_1 = 1$ is indeed odd.

Case II: $k + 1 = 2$. Then $k + 1 \not\equiv 0 \pmod 3$, and $F_2 = 1$ is indeed odd.

Case III: $k + 1 \ge 3$. Then we consider all possible remainders of $k + 1$ modulo 3.

Case IIIA: $k + 1 \equiv 0 \pmod 3$. Then $k \equiv 2 \pmod 3$ and $k - 1 \equiv 1 \pmod 3$. By the inductive hypothesis F_k and F_{k-1} are both odd, so $F_{k+1} = F_k + F_{k-1}$ is even.

Case IIIB: $k + 1 \equiv 1 \pmod 3$. Then $k \equiv 0 \pmod 3$ and $k - 1 \equiv 2 \pmod 3$. By the inductive hypothesis F_k is even and F_{k-1} is odd, so $F_{k+1} = F_k + F_{k-1}$ is odd.

Case IIIC: $k + 1 \equiv 2 \pmod 3$. Then $k \equiv 1 \pmod 3$ and $k - 1 \equiv 0 \pmod 3$. By the inductive hypothesis F_k is odd and F_{k-1} is even, so $F_{k+1} = F_k + F_{k-1}$ is odd.

17.9 Invariants

1. *Proof 1.* We consider all possible cases of signs of the two numbers changed:

   - positive, positive → negative, negative
   - positive, negative → negative, positive
   - negative, negative → positive, positive

We see that the number of positive numbers either does not change or changes by ± 2. Thus the parity of the number of positive numbers is an invariant. We start with the set containing 3 positive numbers. It is not possible to reach 6 positive numbers because 3 is odd but 6 is even.

Proof 2. When two numbers are multiplied by -1, the product of all the numbers does not change. Initially the product is -36. It is not possible to make it 36.

3. An odd number times 3 is an odd number, and an even number times 3 is an even number. So multiplication by 3 does not change the parity of the number. Also, an odd number minus 2 is an odd number, and an even number minus 2 is an even number. So neither of the permitted operations changes the parity of the number. The initial set consists of four odd numbers. Thus the four numbers will always be odd. It is not possible to reach 2 odd and 2 even numbers.

5. When we change the signs of two numbers, the product of all numbers does not change. Initially the product is 1. Since the product of twenty-five -1's is -1, it is not possible to change all numbers into -1.

7. When we replace a by $a + 2b$ or $a - 2b$ (where a and b are two numbers in the set), we do not change its parity (if a is even, then $a \pm 2b$ is even, and if a is odd, then $a \pm 2b$ is odd). Thus the parity of each number will always be the same. Initially we have two even and two odd numbers. It is not possible to make all of the numbers even.

9. The parity of the number of $-$ signs does not change:

   - if two $+$'s are replaced by a $+$, then the number of $-$'s does not change,
   - if two $-$'s are replaced by a $+$, then the number of $-$'s is decreased by 2,
   - if a $+$ and a $-$ are replaced by a $-$, then the number of $-$'s does not change.

   Therefore if we had an even number of $-$ signs, then a $+$ will remain in the end, and if we had an odd number of $-$ signs, then a $-$ will remain in the end.

11. We have seen in problem 3 in chapter 6 that any number is congruent to the sum of its digits modulo 9. Thus the question is equivalent to whether there are more numbers among 1, 2, $\ldots$, 10^6 congruent to 1 or congruent to 2 modulo 9. Remainders of consecutive natural numbers modulo 9 are 1, 2, 3, $\ldots$, 8, 0, and this 9-tuple repeats. The last number in our sequence is $10^6 \equiv 1 \pmod 9$, thus there will be more 1's.

13. When we replace a and b (let $a \geq b$) by $a - b$, the sum of all the numbers changes by

$$-a - b + (a - b) = -2b \equiv 0 \pmod 2.$$

So the parity of the sum does not change. Initially the sum is

$$1 + 2 + \ldots + (4n - 1) = \frac{(4n - 1)4n}{2} = (4n - 1)2n$$

which is even. Thus the sum of the numbers is always even. Therefore an even number will remain in the end.

15. *Proof 1.* The sum of the numbers does not change since a, b, c, d, $\ldots$ are replaced by $2b - a$, $2c - b$, $2d - c$, $\ldots$. The sum of the original numbers is 45. But the sum of ten 5's is 50. Therefore it is not possible to reach ten 5's.

   Proof 2. Since $2b - a \equiv a \pmod 2$, $2c - b \equiv b \pmod 2$, etc., and we start with 5 even and 5 odd numbers, we will always have 5 even and 5 odd numbers. Therefore it is not possible to reach ten 5's.

17. Let the integers in the order they are written be a_1, a_2, a_3, a_4, a_5, and a_6. The sets $\{a_1, a_2, a_3, a_4, a_5, a_6\}$ and $\{1, 2, 3, 4, 5, 6\}$ are equal. Thus the sum of all the a_i's is

$$a_1 + a_2 + a_3 + a_4 + a_5 + a_6 = 1 + 2 + \ldots + 6 = 21.$$

When we add its place number to each integer, we get

$$a_1 + 1, \ a_2 + 2, \ a_3 + 3, \ a_4 + 4, \ a_5 + 5, \ a_6 + 6.$$

The sum of these is

$$(a_1 + 1) + (a_2 + 2) + (a_3 + 3) + (a_4 + 4) + (a_5 + 5) + (a_6 + 6) =$$
$$(a_1 + a_2 + a_3 + a_4 + a_5 + a_6) + (1 + 2 + 3 + 4 + 5 + 6) = 21 + 21 = 42.$$

Suppose that all the sums $a_1 + 1$, $a_2 + 2$, $a_3 + 3$, $a_4 + 4$, $a_5 + 5$, and $a_6 + 6$ have different remainders upon division by 6. Then the remainders are a permutation of the set $\{0, 1, 2, 3, 4, 5\}$ whose sum is

$$0 + 1 + 2 + 3 + 4 + 5 = 15 \equiv 3 \pmod 6.$$

Since $42 \not\equiv 3 \pmod 6$, we get a contradiction.

19. Consider cells with two, three, or four infected neighbors. Notice that when the infection spreads to such a cell, the perimeter of the contaminated area cannot increase (but it may decrease). Namely (look at the picture below), when a cell with two infected neighbors becomes infected, the perimeter of the contaminated area does not change. When a cell with three infected neighbors becomes infected, the perimeter decreases by 2. When a square with four infected neighbors becomes infected, the perimeter decreases by 4. Initially the perimeter is at most $4 \cdot 9 = 36$. It cannot become 40.

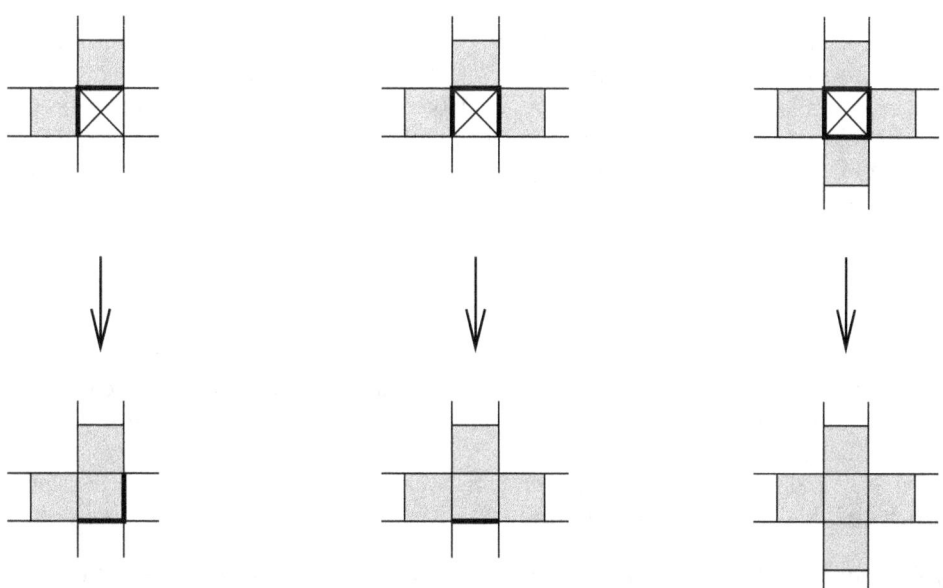

21. First we divide the Parliament into two houses randomly. We will say that a Parliament member is unsatisfied with his placement if he has two or more enemies in his house. If there are unsatisfied members, we will choose any one of them and move him to the other house. Now there is at most one enemy in his house. By this move we reduced the number of hostile pairs (because the member moved was in two hostile pairs and now he is in at most one hostile pair, and no pairs not containing that member were affected by his move). If any unsatisfied members remained, then again we will choose one of them and move him, thus reducing the number of hostile pairs again. And so on. Since it is not possible for the number of hostile pairs to become negative, sooner or later there will be no unsatisfied members.

23. Color the sectors as shown in the picture below.

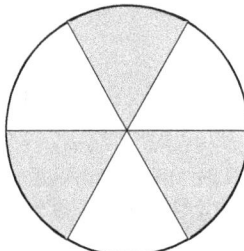

Let S be the sum of the numbers in blue sectors minus the sum of the numbers in white sectors. When we increase the numbers in two neighboring sectors by 1, the quantity S does not change. Initially $S = 2$. If it were possible to equalize all the numbers, S would have to become 0. Therefore it is not possible to equalize all the numbers.

Note. See chapter 10 for more problems where the idea of coloring is helpful.

17.10   Coloring

1. Suppose such a covering is possible. A 14×14 board has 196 squares. Color the board using the standard chessboard coloring. Then it has 98 black squares and 98 white squares. Each T-tetromino covers either 3 black and 1 white or 1 black and 3 white squares. Suppose there are n T-tetrominoes covering 3 black squares. Then there are $49 - n$ T-tetrominoes covering 1 black square. Then all 49 tetrominoes cover $3n + (49 - n) = 2n + 49$ black squares. They must cover 98, so $2n + 49 = 98$, or $2n = 49$. However, this equation has no integer solutions since 49 is not divisible by 2. We get a contradiction.

3. Suppose such a covering is possible. A 10×10 board has 100 squares. Consider the standard chessboard coloring. Then the board has 50 black squares and 50 white squares. Each T-tetromino covers either 3 black and 1 white or 1 black and 3 white squares. Suppose there are n T-tetrominoes covering 3 black squares. Then there are $15 - n$ T-tetrominoes covering 1 black square. Each L-tetromino covers 2 black and 2 white squares. Therefore 10 L-tetrominoes cover 20 black squares. Thus all 25 tetrominoes cover $3n + (15 - n) + 20 = 2n + 35$ black squares. They must cover 50, so $2n + 35 = 50$, or $2n = 15$. However, this equation has no integer solutions since 15 is not divisible by 2. We get a contradiction.

5 Consider the standard chessboard coloring. A T-tetromino covers either 1 or 3 (i.e. an odd number of) black squares. Every other tetromino covers 2 (i.e. an even number of) black squares. If the number of T-tetrominoes were odd, then they would cover an odd number of black squares, and the other tetrominoes would cover an even number of black squares. Thus all tetrominoes together would cover an odd number of black squares. But the chessboard has 32 black squares, and 32 is even. We get a contradiction.

7. There are 36 squares, and each domino covers 2, so we need 18 dominoes. Color the figure as a chessboard (see picture below). It has 20 black and 16 white squares. Since each domino covers one black square and one white square, 18 dominoes must cover 18 black and 18 white squares. Therefore it is not possible to cover the figure with dominoes.

9. Suppose such a covering is possible. Color the board using the stripe pattern with alternating black and white stripes (see the first of the four colorings on page 44). There are 18 black squares. The rest of the argument is the same as in problem 1. Each L-tetromino covers either 1 or 3 black squares. Let n tetrominoes cover 3 black squares, then $9 - n$ tetrominoes cover 1 black square, so all 9 tetrominoes together cover $3n + (9 - n) = 2n + 9$ black squares. Therefore $2n + 9 = 18$, or $2n = 9$. However, this equation has no integer solutions. We get a contradiction.

11. Suppose the upper right corner has been removed.

Proof 1. Color the board diagonally using three colors as shown below.

The board contains 21 white, 22 black, and 20 gray squares. Since each straight tromino must cover 1 white, 1 black, and 1 gray square, the board cannot be covered by straight trominoes.

Proof 2. Suppose such a covering is possible. Color the board using horizontal stripes of three colors, say, from top to bottom: white, blue, black. Then there are 23 white squares (three rows minus one removed square), 24 blue squares (three full rows), and 16 black squares (two full rows). Each tromino covers either three squares of the same color or one square of each color. Let a be the number of trominoes covering three white squares, b the number of trominoes covering three blue squares, c the number of trominoes covering three black squares, and d the number of trominoes covering one square of each color. Then, for the total number of squares of each color, we have: $3a + d = 23$, $3b + d = 24$, and $3c + d = 16$. Subtracting the first equation from the second we have $3b - 3a = 1$, or $3(b - a) = 1$. The left-hand side is divisible by 3 and the right-hand side is not. We get a contradiction.

13. Suppose such a covering is possible. Color the board using the stripe coloring using two colors, say, white and black, starting with black (see the first of the four colorings

on page 44). Then there are 12 black columns and 11 white columns, so there are 23 more black squares than white ones. Each 2×2 tile covers 2 black and 2 white squares, so if there are n of such tiles they cover $2n$ black and $2n$ white squares. Each 3×3 tile covers either 6 or 3 black squares and, respectively, either 3 or 6 white squares. Let m be the number of 3×3 tiles that cover 6 black and 3 white squares, and let k be the number of 3×3 tiles that cover 3 black and 6 white squares. Then the total number of black squares covered by 3×3 tiles is $6m + 3k$, and the total number of white squares covered by 3×3 tiles is $3m + 6k$. Thus the total number of black squares covered by all tiles is $2n + 6m + 3k$, and the total number of white squares covered by all tiles is $2n + 3m + 6k$. Since there are 23 more black squares than white squares, we have $(2n+6m+3k) - (2n+3m+6k) = 23$, or $3m - 3k = 23$, where n, m, and k are integers. However, we see that the left-hand side is divisible by 3, but the right-hand side is not. We get a contradiction.

15. Let a be the number of rows and let b be the number of columns. If $n|a$, then $a = nk$ for some integer k, and each column contains nk squares. Thus we can cover each column by k "vertical" $1 \times n$ tiles. Similarly, if $n|b$, then $b = nk$ for some integer k, and each row contains nk squares. Thus we can cover each row by k "horizontal" $1 \times n$ tiles.

Now suppose that $n \nmid a$ and $n \nmid b$ but an $a \times b$ board can be covered by $1 \times n$ tiles. Color the board diagonally using n colors. Each tile must cover exactly one square of each color. Therefore each color must appear the same number of times. We will show below that this is not possible, thus obtaining a contradiction.

If $a > n$, then in the first n rows each color appears exactly b times (because each color appears exactly once in each column of length n). Therefore if we throw these first n rows away, each color must still appear the same number of times. Similarly, we can throw away the next set of n consecutive rows, and so on, until less than n rows remain. Similarly for the columns. So now we reduced our board to, say, a $c \times d$ board where $c < n$ and $d < n$, and each color must appear the same number of times. Without loss of generality we can assume that $c \leq d$. This $c \times d$ piece is colored diagonally, and we can assign numbers 1 through n to our colors so that they appear in the increasing order as shown in the picture below.

Since $d < n$, the number of colors is at least $d + 1$, so the first $d + 1$ "diagonals" are of different colors. Since only $c - 2 \leq d - 2 < d$ "diagonals" remain, colors d and $d+1$ will not repeat. Therefore in this piece there are c squares of color d but only $c - 1$ squares of color $d + 1$. Thus the colors are not distributed evenly. We get a contradiction.

17. (a) Notice that every piece of a face diagonal connects a vertex and a face midpoint. Thus if we only use face diagonals, vertices and midpoints must alternate. But there are 8 vertices and 6 midpoints, so there is no way to make them alternate (there are too many vertices).

   Note. We could color all the marked points, e.g. let vertices be black, and let midpoints be white. Then black and white points must alternate, but there are

8 black points and 6 white points, so that's impossible. In this problem coloring points was not very useful because it was easy to refer to the points as vertices and face midpoints. In fact, coloring made our solution longer. However, in many problems there is no "natural" division of the points and such a coloring could provide a way to expoain a similar argument.

(b) If one edge is allowed, then we could have two vertices in the beginning, after which we would be left with 6 midpoints and 6 vertices, and we can make them alternate. Again, let vertices be black and midpoints white, then a path could be e.g. *bbwbwbwbwbwbwb*.

Here is an example. (But there are many other such paths.)

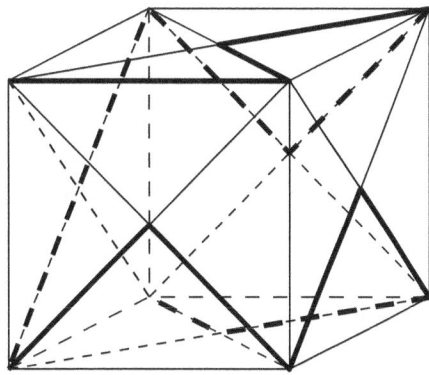

19. Suppose such a filling is possible. "Color" the small (i.e. $1 \times 1 \times 1$) cubes of the $6 \times 6 \times 6$ cube as follows. Color each other "level" as in Example 10.2, and the other "levels" all white.

Then the $6 \times 6 \times 6$ cube contains 27 black cubes. Each $4 \times 1 \times 1$ brick fills either 0 or 2, so, an even number of black cubes. Therefore all bricks together must fill an even number of black cubes. We get a contradiction.

21. No. Assume that such a reentrant knight tour exists. Color the board as shown on the picture below:

Notice that from any black square a knight can only get to a light gray square; from any white square a knight can only get to a dark gray square. Since there are 123 squares of each color, the tour must contain 123 pairs "black, light gray" and 123 pairs "white, dark gray". However, there is no way to get from a light gray square to a white one or from a dark gray square to a black one. We get a contradiction.

23. Hint 1. Use a chessboard-like coloring (namely, two regions that share an edge have different colors).

Hint 2. Consider even and odd numbers.

17.11   Areas and Volumes

1. *Solution 1.* Divide the region into smaller regions whose areas are easy to find, for example:

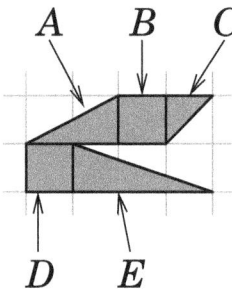

The areas of these regions are: $A = 1$, $B = 1$, $C = \frac{1}{2}$, $D = 1$, $E = \frac{3}{2}$.

Then the total area is the sum of these areas: $1 + 1 + \frac{1}{2} + 1 + \frac{3}{2} = 5$.

Solution 2. Consider the 2×4 rectangle containing the region. Its area is 8. The area of the complement is $F + G + H = 1 + \frac{1}{2} + \frac{3}{2} = 3$, therefore the area of the region is $8 - 3 = 5$.

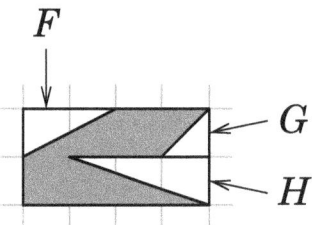

3. Consider the 2×4 rectangle containing the triangle. Its area is 8. The area of the complement is $A + B + C = 1 + 2 + 2 = 5$, therefore the area of the triangle is $8 - 5 = 3$.

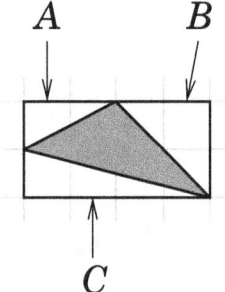

5. Divide the region into two triangles. One of them is a right triangle and has area $A = 1$. By the Pythagorean theorem its hypotenuse is $\sqrt{1^2 + 2^2} = \sqrt{5}$.

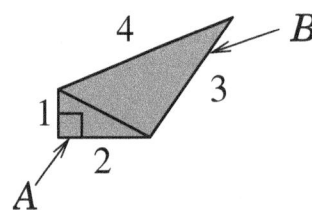

Therefore the area of the second triangle is $B = \sqrt{\frac{p}{2}\left(\frac{p}{2}-a\right)\left(\frac{p}{2}-b\right)\left(\frac{p}{2}-c\right)}$ where a, b, and c are its sides and p is its perimeter: $a = 3$, $b = 4$, $c = \sqrt{5}$, $p = 7 + \sqrt{5}$. Thus we have $B = \sqrt{\frac{7+\sqrt{5}}{2}\left(\frac{7+\sqrt{5}}{2}-3\right)\left(\frac{7+\sqrt{5}}{2}-4\right)\left(\frac{7+\sqrt{5}}{2}-\sqrt{5}\right)} =$

$\sqrt{\frac{7+\sqrt{5}}{2}\cdot\frac{1+\sqrt{5}}{2}\cdot\frac{\sqrt{5}-1}{2}\cdot\frac{7-\sqrt{5}}{2}} = \sqrt{\frac{(7+\sqrt{5})(7-\sqrt{5})(\sqrt{5}+1)(\sqrt{5}-1)}{16}} =$

$\sqrt{\frac{(49-5)(5-1)}{16}} = \sqrt{\frac{44\cdot 4}{16}} = \sqrt{11}$.

Therefore the total area is $A + B = 1 + \sqrt{11}$.

7. Draw two heights of the trapezoid as shown in the picture below.

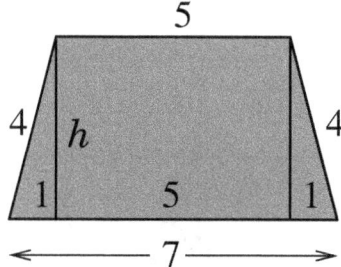

Since the two triangles on the sides are congruent, the heights divide the base into segments of length 1, 5, and 1. Then the height is $h = \sqrt{4^2 - 1^2} = \sqrt{15}$, therefore the area of the trapezoid is $A = \frac{1}{2}(5 + 7)\sqrt{15} = 6\sqrt{15}$.

9. Consider the 2×2 square containing the region. Its area is 4. The complement of the region in the square consists of four quarters of a circle with radius 1, therefore the total area of the complement is the same as the area of the circle, i.e. π. Therefore the area of the region is $4 - \pi$.

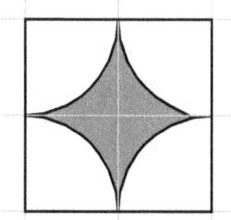

11. Consider the following sectors:

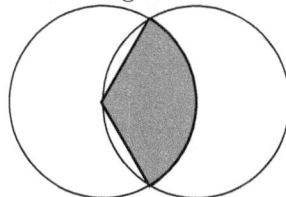 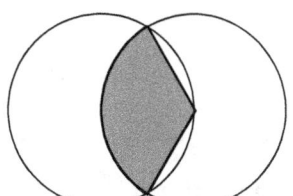

Each of the above sectors is one-third of a full circle of radius 1, thus has area $\frac{\pi}{3}$. Their total area is $\frac{2\pi}{3}$, and they cover our region, but they overlap. We must subtract the area of the overlap from $\dfrac{2\pi}{3}$ to obtain the area of the region. The overlap consists of two equilateral triangles with side 1:

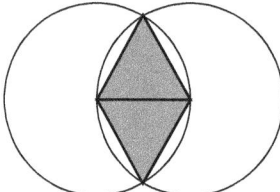

Using the Pythagorean theorem it is easy to find that the height of such a triangle is $\dfrac{\sqrt{3}}{2}$, and thus the area is $\dfrac{\sqrt{3}}{4}$. The total area of the two triangles is then $\dfrac{\sqrt{3}}{2}$, and the area of the original region is $\dfrac{2\pi}{3} - \dfrac{\sqrt{3}}{2}$.

13. Divide the region into ten regions (four equilateral triangles and 6 thin pieces around them) as shown in the picture below.

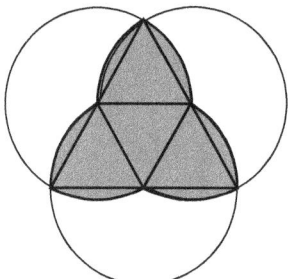

The area of each triangle is $\dfrac{\sqrt{3}}{4}$ (see problem 11). The area of each thin piece is $\dfrac{\pi}{6} - \dfrac{\sqrt{3}}{4}$ (the area of one-sixth of the circle minus the area of the triangle). Thus the total area of the region is $4 \cdot \dfrac{\sqrt{3}}{4} + 6\left(\dfrac{\pi}{6} - \dfrac{\sqrt{3}}{4}\right) = \pi - \dfrac{\sqrt{3}}{2}$.

15. Consider one of the four leaves in the figure. It is the overlap of two sectors, one quarter of a circle (of radius $\dfrac{1}{2}$) each. The area of a quarter of a circle is $\dfrac{\pi}{16}$, therefore the area of the overlap is $2 \cdot \dfrac{\pi}{16} - \dfrac{1}{4} = \dfrac{\pi}{8} - \dfrac{1}{4}$.

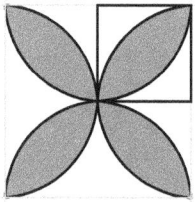

The total area of the region is $4\left(\dfrac{\pi}{8} - \dfrac{1}{4}\right) = \dfrac{\pi}{2} - 1$.

17. Since each of x and y can be nonnegative or negative, we consider the following four cases.

Case I: $x \geq 0$, $y \geq 0$ (first quadrant). Then $|x| = x$ and $|y| = y$. The inequality becomes $x + 2y \leq 4$, or $y \leq -\dfrac{x}{2} + 2$.

Case II: $x \geq 0$, $y < 0$ (fourth quadrant). Then $|x| = x$ and $|y| = -y$. The inequality becomes $x - 2y \leq 4$, or $y \geq \dfrac{x}{2} - 2$.

Case III: $x < 0$, $y \geq 0$ (second quadrant). Then $|x| = -x$ and $|y| = y$. The inequality becomes $-x + 2y \leq 4$, or $y \leq \dfrac{x}{2} + 2$.

Case IV: $x < 0$, $y < 0$ (third quadrant). Then $|x| = -x$ and $|y| = -y$. The inequality becomes $-x - 2y \leq 4$, or $y \geq -\dfrac{x}{2} - 2$.

Now we draw the region in each case and obtain the following figure.

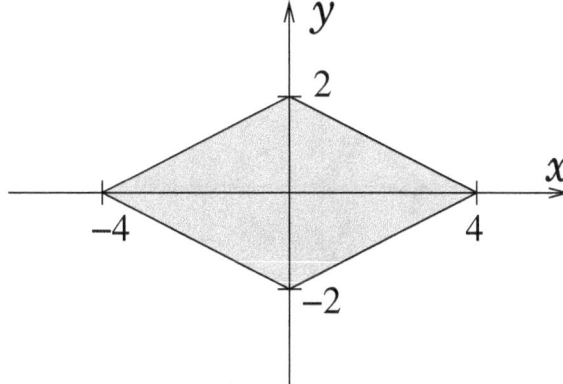

This figure consists of four triangles with base 4 and height 2, therefore its area is $4 \cdot \dfrac{1}{2} \cdot 4 \cdot 2 = 16$.

19. Clearly, the base of the box is a square. Let the base be a cm $\times$ a cm. Since the height of the box is 3 cm, its volume is $3a^2$ cubic cm. We are given that the volume is 60 cubic cm, therefore we have $3a^2 = 60$, so $a^2 = 20$, and $a = \sqrt{20} = 2\sqrt{5}$. Thus the box has dimensions $2\sqrt{5}$ cm $\times$ $2\sqrt{5}$ cm $\times$ 3 cm.

21. Consider the top half of the octahedron.

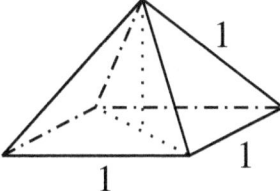

It is a pyramid with a square 1×1 base (whose area is 1). The base diagonal has length $\sqrt{2}$, therefore the distance from the center of the octahedron to any vertex is $\dfrac{\sqrt{2}}{2}$. Thus the height of the pyramid is $\dfrac{\sqrt{2}}{2}$, and then its volume is $\dfrac{1}{3} \cdot 1 \cdot \dfrac{\sqrt{2}}{2} = \dfrac{\sqrt{2}}{6}$. Since the octahedron consists of two such pyramids, its total volume is $2 \cdot \dfrac{\sqrt{2}}{6} = \dfrac{\sqrt{2}}{3}$.

23. Let a be half the length of any edge of the cube. By the Pythagorean theorem, the distance from the center of the cube to any of its vertices is $a\sqrt{3}$. On the other hand,

this distance is the radius of the sphere, so $a\sqrt{3} = 1$. Therefore $a = \dfrac{1}{\sqrt{3}}$, the edge has

length $\dfrac{2}{\sqrt{3}}$, and thus the volume of the cube is $\dfrac{8}{3\sqrt{3}}$.

Next let us introduce the coordinate system so that the center of the cube and the sphere is at the origin and the z-axis is vertical as shown in the picture below. We will use an integral to calculate the volume of the solid inside the sphere and above the plane $z = \dfrac{1}{\sqrt{3}}$ as follows: the horizontal cross-section at z is a disk with radius $r = \sqrt{1 - z^2}$

(because by the Pythagorean theorem $z^2 + r^2 = 1$), therefore $V = \displaystyle\int_{\frac{1}{\sqrt{3}}}^{1} \pi(\sqrt{1 - z^2})^2 dz =$

$$\pi \int_{\frac{1}{\sqrt{3}}}^{1} (1 - z^2)dz = \pi \left(z - \frac{z^3}{3} \right)\Big|_{\frac{1}{\sqrt{3}}}^{1} = \left(1 - \frac{1}{3} - \frac{1}{\sqrt{3}} + \frac{1}{9\sqrt{3}} \right)\pi = \left(\frac{2}{3} - \frac{8}{9\sqrt{3}} \right)\pi.$$

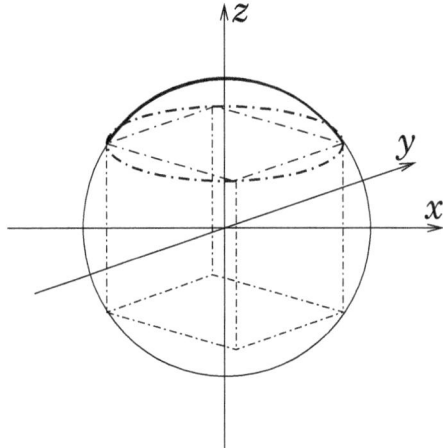

Notice that the cube and six "caps", one adjacent to each side of the cube, whose volume we calculated above, fill the cube with the overlap consisting of twelve thin pieces, one along each edge of the cube.

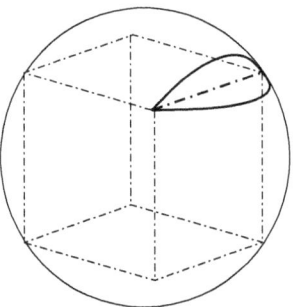

Since the volume of the sphere is $\dfrac{4}{3}\pi$, the total volume of these twelve thin pieces is

$\dfrac{8}{3\sqrt{3}} + 6\pi \left(\dfrac{2}{3} - \dfrac{8}{9\sqrt{3}} \right) - \dfrac{4}{3}\pi = \dfrac{8}{3\sqrt{3}} + \dfrac{8}{3}\pi - \dfrac{16}{3\sqrt{3}}\pi$. The volume of the solid inside the sphere and above the cube is the volume of one "cap" minus the volume of four thin

pieces, i.e. $\left(\dfrac{2}{3} - \dfrac{8}{9\sqrt{3}} \right)\pi - \dfrac{4}{12}\left(\dfrac{8}{3\sqrt{3}} + \dfrac{8}{3}\pi - \dfrac{16}{3\sqrt{3}}\pi \right) = \dfrac{8}{9\sqrt{3}}\pi - \dfrac{8}{9\sqrt{3}} - \dfrac{2}{9}\pi.$

17.12 Symmetry, Translations, Rotations, and Similarity

1. Assume that a solution exists. Rotate circle S through an angle of 30° around A so that the image of B is C. Let S' be the image of S. Then C is an intersection point of S' and T.

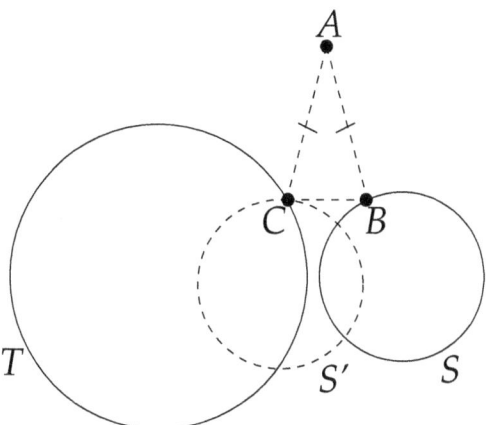

Thus to find a solution, we rotate circle S through an angle of 30° around A (say, in the clockwise direction; if its image does not intersect circle T, then rotate S in the counterclockwise direction). Let S' be its image, and let C be an intersection point of S' and T. Rotate the point C back – let B be its image. It lies on the original circle S. Then $AB = AC$ and $\angle BAC = 30°$. The other angles are as required because $\angle ABC = \angle ACB$ since $\triangle ABC$ is isosceles, and the sum of all the angles in a triangle is 180°.

3. (a) Let l be the given line.

   Case I: points A and C lie on the opposite sides of l. Since the shortest path from A to C is a straight line, draw a line through these points, and let B be the intersection point of AC and l.

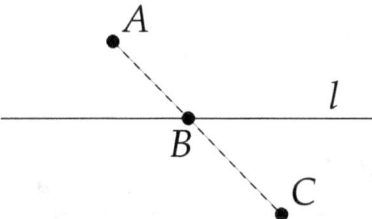

   Case II: points A and C lie on the same side of l. Reflect point C about line l, and let C' be its image. Since for any point B on l we have $|BC| = |BC'|$, minimizing $|AB| + |BC|$ is equivalent to minimizing $|AB| + |BC'|$. The latter is minimal when A, B, and C' lie on one line.

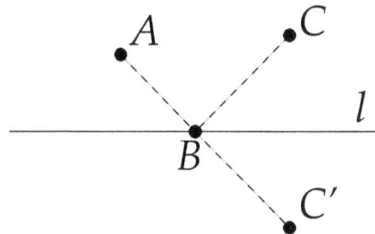

Case III: at least one of points A and C lies on l. Without loss of generality we can assume that point A lies on l. Since the shortest path from A to C is a straight line, taking $B = A$ minimizes $|AB| + |BC|$.

5. Assume that a solution exists. Draw a vertical line l through A. Reflect line p about l, let p' be its image. Let I be the intersection point of l and BC. Since BC is horizontal, $\angle AIB = \angle AIC$. Since also $|AB| = |AC|$, triangles AIB and AIC are congruent. Therefore the image of B is C. Thus p' and q have point C in common.

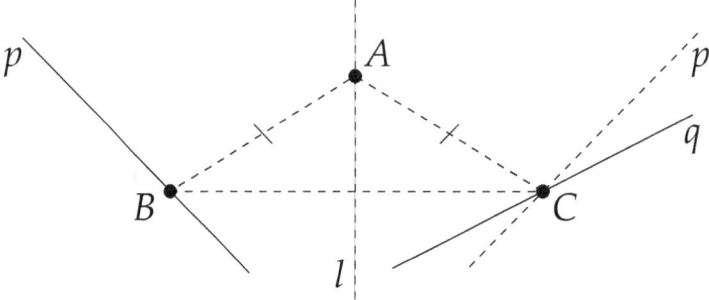

Thus to find a solution, draw a vertical line l through A. Reflect line p about l, let p' be its image. Let C be the intersection point of p' and q. Now reflect the point C about l, and let B be its image. It lies on p. Since B and C are symmetric about a vertical line l, we have that BC is horizontal and $|AB| = |AC|$.

7. Draw any angle that is not a multiple of π and mark segments of lengths a, b, and c on its sides as shown in the picture below.

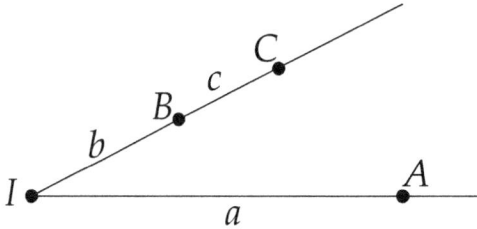

Draw a line through points A and C, and then draw a line through B and parallel to AC. Let D be the intersection point of this line with the other side of the angle, and let $x = |ID|$ and $y = |DA|$.

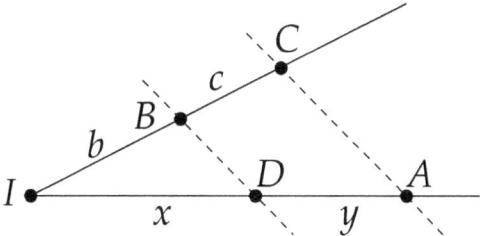

Since $\triangle ICA$ and $\triangle IBD$ are similar, we have $\dfrac{x+y}{x} = \dfrac{b+c}{b}$, or $1 + \dfrac{y}{x} = 1 + \dfrac{c}{b}$. Therefore $\dfrac{y}{x} = \dfrac{c}{b}$, so $\dfrac{x}{y} = \dfrac{b}{c}$.

9. (a) Let S and T be the centers of the circles, and let r and r' be their radii respectively. If $r = r'$, then a common tangent line is parallel to line ST, thus it suffices to draw radii perpendicular to ST and draw a line through the ends of the radii.

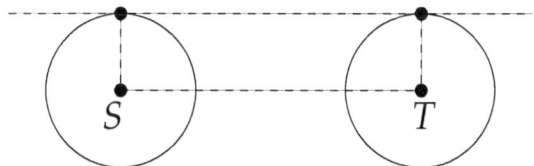

If $r \neq r'$, then without loss of generality we can assume that $r > r'$.

Solution 1. Draw a circle centered at S with radius $r - r'$. Then draw a tangent line to this circle that passes through T. This can be done e.g. by drawing a semi-circle with diameter ST. If A is the intersection point of the semicircle and the circle with radius $r - r'$, then $\angle SAT = 90°$.

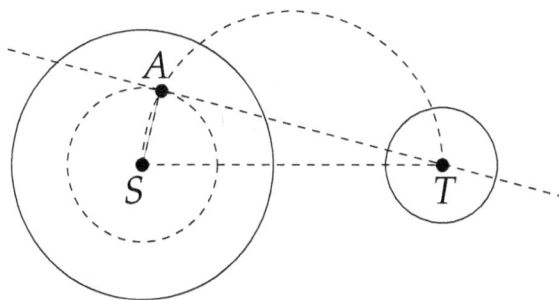

Extend SA until it intersects the given circle centered at S, and let B be this intersection point. Let TC be the radius of the second given circle perpendicular to AT so that B and C lie on the same side of AT. Then $|AB| = |CT| = r'$, $\angle BAT = \angle ATC = 90°$, therefore $ABCT$ is a rectangle. Therefore BC is perpendicular to both SB and TC, so it is tangent to both of the given circles.

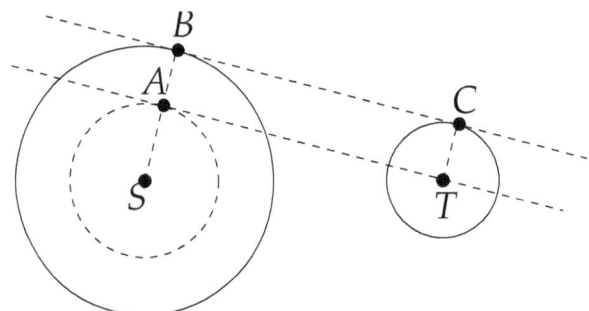

Solution 2. Let $d = |ST|$. Draw a line through S and T. It must cross the common tangent line that we are looking for. Let us find the location of the intersection point I. Let $x = |TI|$, then $|SI| = d + x$.

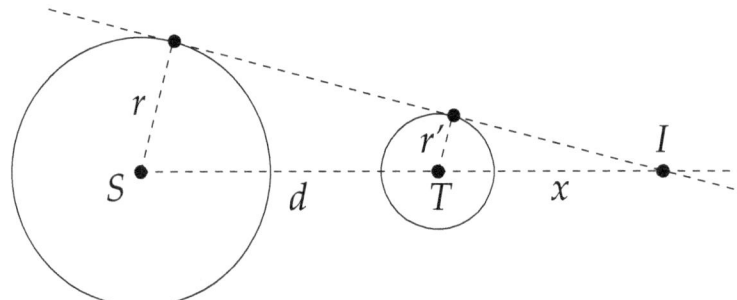

From similar triangles we see that $\dfrac{x}{r'} = \dfrac{d+x}{r}$. Solving this equation for x gives $x = \dfrac{r'd}{r - r'}$. See problem 6 about constructing a segment of this length.

Once we have this intersection point, we draw semicircles with diameters SI and TI, and find the intersection points A and B of these semicircles with the given circles. Since $\angle SAI = \angle TBI = 90°$, line AB is tangent to both circles.

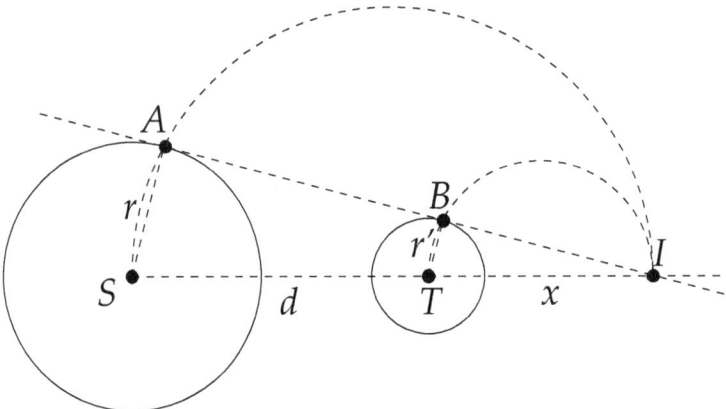

11. *Solution 1.* Since $\angle CAB = 60°$ and AD bisects $\angle CAB$, $\angle DAB = 30°$. Then $\angle ADB = 60°$. Thus the problem is equivalent to finding points A and D such that $\angle ADB = 60°$ and $|AD| = l$. To do this, translate line q a distance l in the direction of the ray r such that the angle between r and q is $60°$ (see picture below). Let q' be its image. Let A be the intersection point of p and q'. Translate A back – get point D on the original line q. Now draw a vertical line v through A. Let B be the intersection point of v and

q. By our construction, $|AD| = l$, AB is vertical, $\angle ADB = 60°$, thus $\angle DAB = 30°$, and thus AD bisects $\angle CAB$.

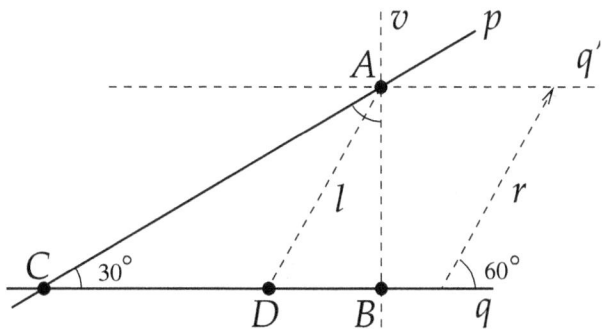

Solution 2. Since $\angle DAB = 30°$, $\dfrac{|AB|}{|AD|} = \cos 30° = \dfrac{\sqrt{3}}{2}$. Then $|AB| = \dfrac{\sqrt{3}}{2}|AD| = \dfrac{\sqrt{3}}{2}l$.

Thus we have to translate q a distance $\dfrac{\sqrt{3}}{2}l$ upward (this can be done with a ruler and a compass), and let A be the intersection point of the new line q' and p. (Note: the new line q' here is, of course, the same as in solution 1.) Then draw a vertical line through A to find B.

Solution 3. As in solution 2, $|AB| = \dfrac{\sqrt{3}}{2}l$. Now, since $\angle ACB = 30°$, $\dfrac{|AB|}{|CB|} = \tan 30° = \dfrac{1}{\sqrt{3}}$. Then $|CB| = \sqrt{3}|AB| = \dfrac{3}{2}l$. Thus we merely have to find B on q such that $|CB| = \dfrac{3}{2}l$, and then draw a vertical line through B to find A.

Solution 4. Since we need $\angle CAD = 30° = \angle DCA$, $\triangle CAD$ must be isosceles. Thus we need $|CD| = |AD| = l$. Therefore, we have to find D on q such that $|CD| = l$, then draw a line AD such that $\angle ADC = 120°$, and, finally, draw the vertical line AB.

Solution 5. Choose any point A' on p. Draw a vertical line v' through A', and let B' be the intersection point of v' and q. Draw the bisector $A'D'$ of $\angle CA'B'$. The length of $A'D'$ is most likely not equal to l (if it is, then we are done). So we have to adjust our construction (proportionally) to make this distance equal to l. Namely, let $|A'D'| = l'$. Since we need $|AD| = l$, $\triangle CAB$ we are looking for is similar to $\triangle CA'B'$ with coefficient of similarity $\dfrac{l}{l'}$. Thus, we find A on p such that $|CA| = \dfrac{l}{l'}|CA'|$. See problem 6 on how construct a segment of this length. Then draw a vertical line through A to find B as before, and, finally, draw the bisector AD. From similar triangles, we have $\dfrac{|AD|}{|A'D'|} = \dfrac{|CA|}{|CA'|} = \dfrac{l}{l'}$. Then $|AD| = \dfrac{l}{l'}|A'D'| = \dfrac{l}{l'}l' = l$ as desired.

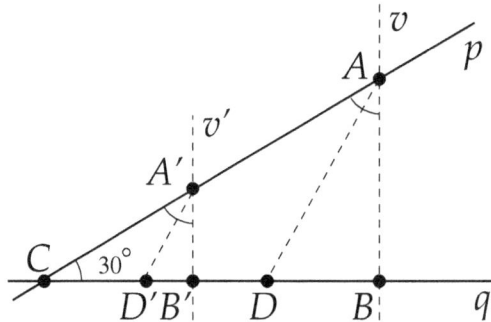

Note. Solution 4 works only because $\triangle ACD$ must be isosceles which uses the fact that the given angle is $30°$ and AB must be vertical. Solutions 2 and 3 may work for some other angles and directions of AB, but they still use the fact that all the angles are "nice". Solutions 1 and 5 would work for any angle and any direction of AB.

Note. The idea of solution 5 is very useful for many problems.

13. Assume that a solution exists. Notice that $|AC| = \sqrt{2}\,|AB|$ and $\angle BAC = 45°$. This means that if we rotate point B through $45°$ around A, then the image B' lies on line AC and $|AC| = \sqrt{2}\,|AB'|$.

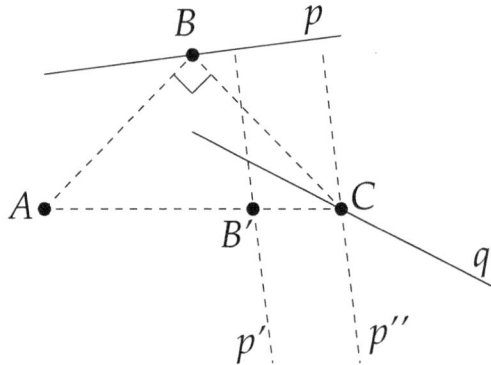

Thus to find such points, we will rotate line p through an angle of $45°$ around A. Let p' be the image. Then we draw a line p'' parallel to p' and such that the distance from A to p'' is $\sqrt{2}$ times the distance from A to p'. Let C be the intersection point of p'' and q. (If p'' and q do not intersect, then rotate p in the opposite direction to find p'.) Draw a line through A and C. Let B' be the intersection point of AC and p'. Rotate B' through an angle of $45°$ around A (in the opposite direction than we rotated line p'). Let B be its image. Since $|AB| = |AB'| = \dfrac{|AC|}{\sqrt{2}}$ and $\angle BAC = 45°$, triangle ABC is as required.

15. Rotate the lower half of the left small circle through $45°$, $90°$, and $135°$ in the clockwise direction around the center of the bigger circle. Its images cut the region into 4 congruent parts.

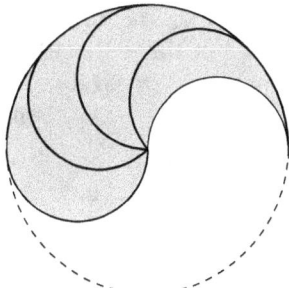

16. Hint: choose any point on the first line as one of the vertices.

17. Reflect the point A about p, and let A' be its image. Then reflect A about q, and let A'' be its image. Since $|AB| + |BC| + |CA| = |A'B| + |BC| + |CA''|$, minimizing the perimeter of triangle ABC is equivalent to minimizing $|A'B| + |BC| + |CA''|$. The latter is minimized when A', B, C, and A'' lie on one line (because the shortest path from A' to A'' is a straight line). Thus we connect A' and A'', and let B and C be the intersection points of $A'A''$ with lines p and q respectively.

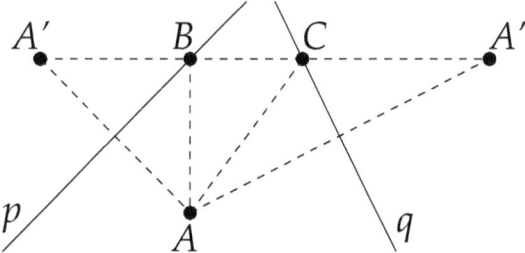

19. Case I: lines r and s are parallel. Then for any horizontal line intersecting r and s at C and D respectively, $|CD|$ is the same. Let $m = l - |CD|$. Thus we have to construct a horizontal line intersecting p and q at A and B such that $|AB| = m$. Assume that a solution exists. Move line p a distance m in a horizontal direction. Let p' be its image. Then p' and q have point B in common.

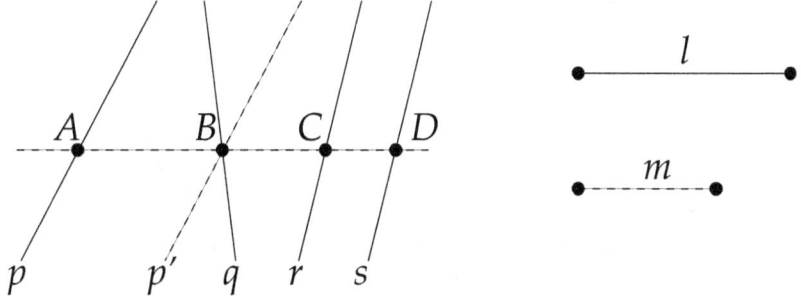

Thus to find a solution, we have to move line p a distance m in a horizontal direction and let B be the intersection point of the image and q. Then draw a horizontal line through B and let A, C, and D be the intersection points as described in the problem.

Case II: lines r and s are not parallel. Let I be their intersection point. First we will construct a line t such that for any horizontal line intersecting p, q, r, s, and t at points A, B, C, D, and E respectively, $|AB| + |CD| = |AE|$. Draw a horizontal line through I and let J be the intersection point of this line and q. Next choose any point K on s that lies to the right of r, and draw a horizontal line through K. Let L and M be the intersection points of this line with q and r respectively. Construct the point N on line LK such that $|LN| = |MK|$. Let t be the line through J and N. Then for horizontal line intersecting p, q, r, s, and t at points A, B, C, D, and E respectively, $\triangle JBE$ and $\triangle JLN$ are similar, and $\triangle ICD$ and $\triangle IMK$ are similar. Therefore $\dfrac{|BE|}{|LN|} = \dfrac{|JB|}{|BL|} = \dfrac{|IC|}{|CM|} = \dfrac{|CD|}{|MK|}$. Since $|LN| = |MK|$, we have $|BE| = |CD|$, therefore $|AB| + |CD| = |AB| + |BE| = |AE|$. Now we construct a horizontal line intersecting p and t at points A and E such that $|AE| = l$. If p and t are parallel, then $|AE|$ must be equal to l for any such line in order for a solution to exist. If p and t are not parallel, then the construction is similar to case I.

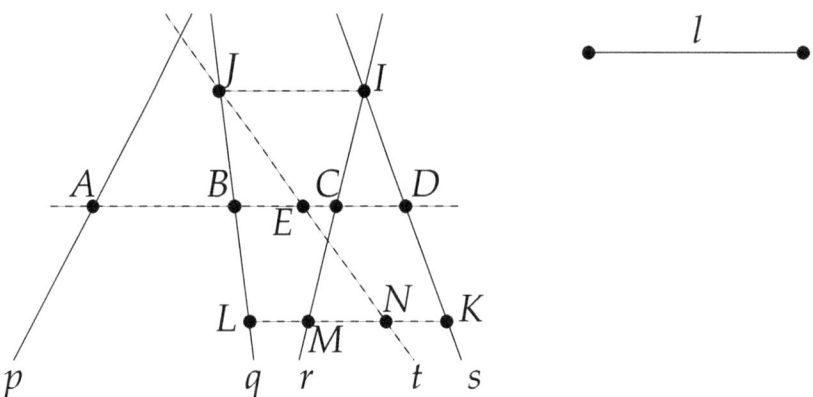

17.13   Graphs

1. By corollary 13.6, in any graph, the number of vertices of odd degree is even. Here there are 3 vertices of degree 3, so there is no such graph.

3. (a) Yes. Here are two examples.

   (b) No. If one vertex has degree 7, then it is connected with every other vertex. Hence there cannot be a vertex of degree 0.

5. Let vertices represent people, and edges represent friendship (two vertices are connected if and only if the corresponding people are friends). Then the degree of each vertex is the number of friends of the corresponding person. Since in any graph the number of vertices of odd degree is even, we have that the number of people with an odd number of friends is even.

7. We can start with any vertex and assume it's in set X. Then consider any vertex connected with the first one, and if the graph is bipartite, this vertex must be in the

other set, say, Y. Then consider any vertex connected with the first or second one, and so on. If we ever run into a situation when two vertices in one set are connected, the graph is not bipartite. If not, we'll have a division of the set of vertices into two sets X and Y such that there are no edges within one set, and hence the graph is bipartite.

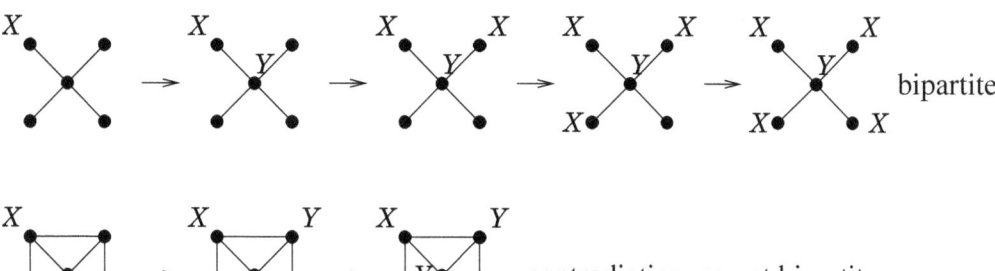

bipartite

contradiction, so not bipartite

Similarly:

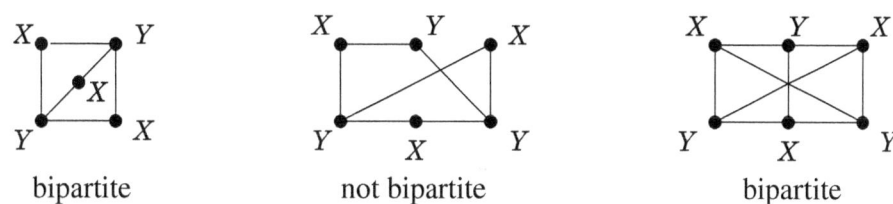

bipartite   not bipartite   bipartite

9. If this were possible, consider the following graph with 8 vertices: each vertex represents a county, and two vertices are connected if and only if the corresponding counties are neighbors. Then the degree of each vertex is the number of the neighbors of that county. Thus we would have a graph with 8 vertices of degrees 5, 5, 4, 4, 4, 4, 4, 3. But in any graph, the sum of the degrees of all the vertices is even. The sum $5+5+4+4+4+4+4+3 = 33$ is odd. Contradiction.

11. A graph has an Euler path but not an Euler cycle iff it is connected and has exactly two vertices of odd degree. The fact that it does not have an Euler cycle follows from Theorem 13.14. Here is a proof that it contains an Euler path. Connect the two vertices of odd degrees. Then all degrees become even and thus there exists an Euler cycle. Now remove this added edge and obtain an Euler path of the original graph.

13. First of all, recall that $K_{n,m}$ has 2 groups of vertices, n vertices in group A, m vertices in group B, and every vertex in group A is connected to every vertex in group B.

   (a) We know that an Euler cycle exists if and only if the degree of each vertex is even. The graph $K_{n,m}$ has n vertices of degree m and m vertices of degree n. Thus an Euler cycle exists iff both m and n are even.

   (b) By problem 11, an Euler path exists if and only if the graph has at most 2 vertices of odd degree. We have the following cases:

   (1) No vertices of odd degree, i.e. all the degrees are even. Then both m and n are even.

   (2) There are two vertices of odd degree and they are in the same group, say, in group A of n vertices. Since all the vertices in this group have the same (odd) degree, and we can have at most 2 vertices of odd degree, there are only 2 vertices

in this group, thus $n = 2$. Since their degree is odd, m is odd. Thus we have $n = 2$ and m is odd.

(3) There are two vertices of odd degree, one in group A and the other in group B. Then both m and n are odd, thus all the degrees are odd, but we can have at most 2 odd degrees, so $n = m = 1$.

(c) A Hamilton cycle is a cycle that visits every vertex exactly once. If a Hamilton cycle starts at a vertex in group A, then its second vertex belongs to group B, the next one belongs to group A, the fourth one belongs to group B, and so on, i.e. A and B will alternate. It must eventually come back to the original vertex, therefore the number of vertices in group A must be equal to the number of vertices in group B. Thus $m = n$. Conversely, if $m = n$, then a Hamiliton cycle can be found by alternating groups A and B.

(d) A path does not return to the starting point, thus in addition to the case $m = n$ (in this case a path has the form ABAB...AB), we have $m = n - 1$ (then we can find a path of the form ABAB...ABA), and $m = n + 1$ (then we can find a path of the form BABA...BAB).

15. First draw the graph representing all possible moves of a knight:

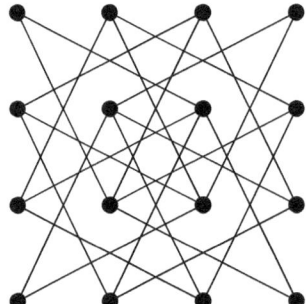

A reentrant tour is a Hamilton cycle. Thus we have to show that this graph has no Hamilton cycle. Notice that there are 4 vertices of degree 2, and in order to visit a vertex of degree 2 we have to use both its edges. Consider the upper left corner vertex and the lower right corner vertex. We must use both edges at each of them. But then we get a cycle. There is no way of adding anything to this cycle (because if we add more edges, we'll have to go through some vertex more than once). But this cycle misses many vertices. Thus there is no Hamilton cycle.

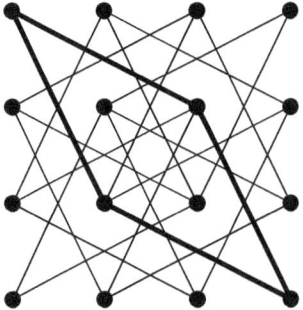

17. If nobody made a mistake, we would be able to draw a bipartite graph with 14 vertices, 7 vertices representing men and 7 vertices representing women, with 2 vertices connected

if and only if the corresponding people shared a dance. Then the sum of the degrees of the 7 vertices representing men should be equal to the sum of the degrees of the 7 vertices representing women (both sums being equal to the number of edges). But it is not possible to divide the given 14 numbers into 2 groups such that the sums are equal because one group must contain the 5, and the other group must consist of 3's and 6's. The sum of the numbers in the first group is congruent to 2 modulo 3, and the sum of the numbers in the second group is congruent to 0 modulo 3, so the sums cannot be equal.

19. Let scientists be represented by vertices. Connect all vertices. We get the complete graph K_{17}. Color the edges using three colors, say, blue, red, and green, according to the topic discussed by the scientists these vertices represent. We have to prove that there are three vertices connected (pairwise) by 3 edges of the same color. Choose any vertex, say, vertex A. It is connected with 16 other vertices. Among the 16 edges connecting vertex A with other vertices at least 6 are of the same color, say, blue. Look at those 6 vertices. If at least two of them are connected by a blue edge then we have a blue triangle. If not, look at the K_6 graph for those 6 vertices. All its edges are red and green. By example 13.18, it contains at least one triangle with all 3 sides of the same color, either red or green.

21. (a) Yes. See the picture below.

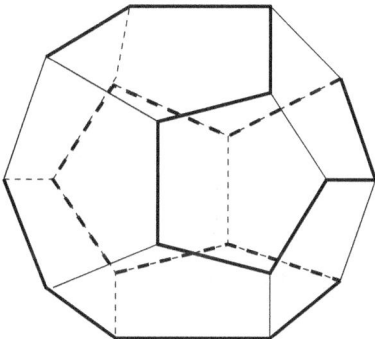

(b) Yes. In the above picture, we can connect the two ends of the path.

23. Let the degrees of the remaining vertices be a (in group A) and b (in group B). The sum of degrees of vertices in the first group must be equal to the sum of degrees of vertices in the second group. Thus $4 + 2 + 2 + a = 3 + 1 + 1 + b$, or $3 + a = b$. Since the graph is connected, the degree of each vertex is at least 1. Thus $a \geq 1$. Now, it is easy to see that for every pair a, b satisfying $3 + a = b$ and $a \geq 1$, there exists a graph with vertices of such degrees. (Because we can have multiple edges between the vertices of degrees a and b.) Draw a few such graphs!

25. The graph $K_{1,2,4}$ is shown below.

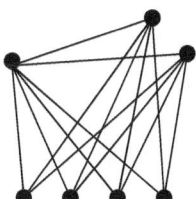

(a) The degrees of the vertices are 6, 5, 5, 3, 3, 3, 3. Since there are vertices of odd degree, there is no Euler cycle.

(c) Yes. See picture below.

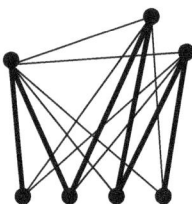

27. We will use a graph to represent the city as follows. Let each of the four pieces of land be represented by a vertex, and let each bridge be represented by an edge connecting the corresponding vertices. Then we get the following graph:

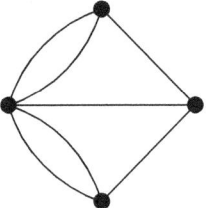

Existence of a tour described in the problem is equivalent to existence of an Euler cycle in this graph. We know that an Euler cycle exists if and only if the degree of each vertex is even. However, the degrees of the vertices in our graph are 5, 3, 3, and 3, i.e. all odd, and this is how we know that no such tour exists. Now, if we add one more bridge we will change the degrees of two vertices (corresponding to the pieces of land that this bridge connects), thus the degrees of two vertices will become even, but the degrees of the other two vertices will remain odd. Therefore no such tour will exist even if one bridge is built.

17.14   Working backwards

1. (a) $46 = 1 \cdot 32 + 14$
   $32 = 2 \cdot 14 + 4$
   $14 = 3 \cdot 4 + 2$
   $4 = 2 \cdot 2$
   Thus $d = (46, 32) = 2$.
   $$\begin{aligned} 2 &= 14 - 3 \cdot 4 \\ &= 14 - 3(32 - 2 \cdot 14) = 7 \cdot 14 - 3 \cdot 32 \\ &= 7(46 - 1 \cdot 32) - 3 \cdot 32 = 7 \cdot 46 - 10 \cdot 32 \end{aligned}$$
   Thus $x = 7$ and $y = -10$.

   (c) $96 = 1 \cdot 54 + 42$
   $54 = 1 \cdot 42 + 12$
   $42 = 3 \cdot 12 + 6$
   $12 = 2 \cdot 6$
   Thus $d = (96, 54) = 6$.

$$6 = 42 - 3 \cdot 12$$
$$= 42 - 3(54 - 1 \cdot 42) = 4 \cdot 42 - 3 \cdot 54$$
$$= 4(96 - 1 \cdot 54) - 3 \cdot 54 = 4 \cdot 96 - 7 \cdot 54, \text{ so } x = 4 \text{ and } y = -7.$$

3. Choose any remainder r_7 and quotients q_1 through q_8. E.g. let all of these be equal to 2, then
$r_6 = q_8 \cdot r_7 = 2 \cdot 2 = 4$,
$r_5 = q_7 \cdot r_6 + r_7 = 2 \cdot 4 + 2 = 10$,
$r_4 = q_6 \cdot r_5 + r_6 = 2 \cdot 10 + 4 = 24$,
$r_3 = q_5 \cdot r_4 + r_5 = 2 \cdot 24 + 10 = 58$,
$r_2 = q_4 \cdot r_3 + r_4 = 2 \cdot 58 + 24 = 140$,
$r_1 = q_3 \cdot r_2 + r_3 = 2 \cdot 140 + 58 = 338$,
$b = q_2 \cdot r_1 + r_2 = 2 \cdot 338 + 140 = 816$,
$a = q_1 \cdot b + r_1 = 2 \cdot 816 + 338 = 1970$.
Then reversing the order of the above equations gives divisions in Euclid's algorithm.

7. Reflect the given graph about the x-axis (i.e. multiply the function by -1) and shift 3 units upward (i.e. add 3).

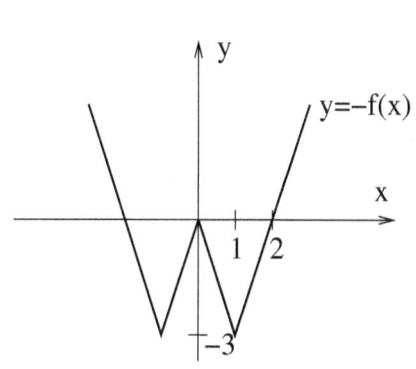 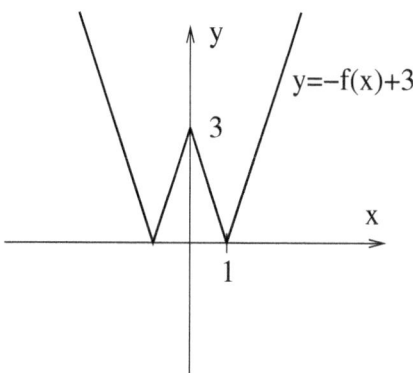

Then $-f(x) + 3 = |g(x)|$ where

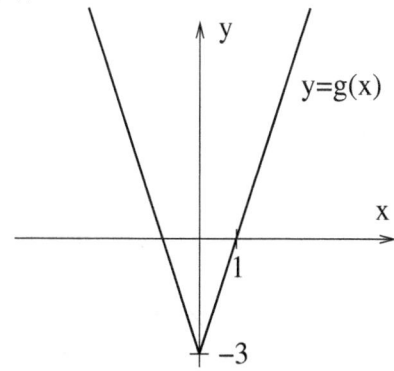

Shifting the graph of $g(x)$ 3 units upward will give the graph of $|3x|$, therefore

$g(x) + 3 = |3x|$
$g(x) = |3x| - 3$
$-f(x) + 3 = ||3x| - 3|$
$f(x) = 3 - ||3x| - 3|$

9. Let's denote the function whose graph is given by $f(x)$. Let $g(x) = f(x) - \dfrac{x}{2}$ (using the

hint given in problem 8). It's easiest to sketch the graph of $g(x)$ if we write piece-wise linear formulas for $f(x)$ and $g(x)$ first:

$$f(x) = \begin{cases} 0 & \text{if } x < 0 \\ x & \text{if } 0 \le x < 1 \\ 2x - 1 & \text{if } x \ge 1 \end{cases}, \quad \text{so} \quad g(x) = \begin{cases} -x/2 & \text{if } x < 0 \\ x/2 & \text{if } 0 \le x < 1 \\ 3x/2 - 1 & \text{if } x \ge 1 \end{cases}.$$

Then the graphs are as follows:

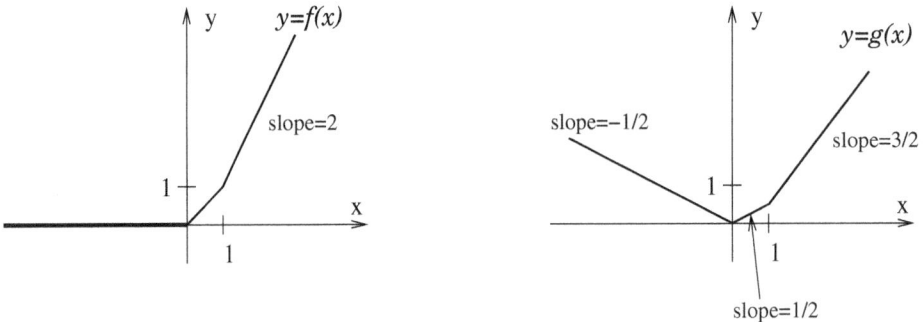

We see that $g(x) = |h(x)|$ where the graph of $h(x)$ is:

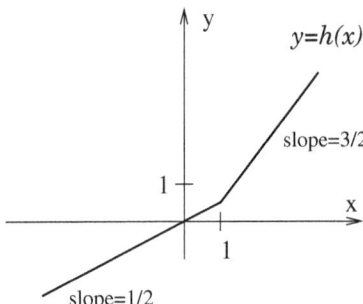

Now we will move the graph of $h(x)$ so that the vertex is at the origin, let's call the new function $k(x)$:

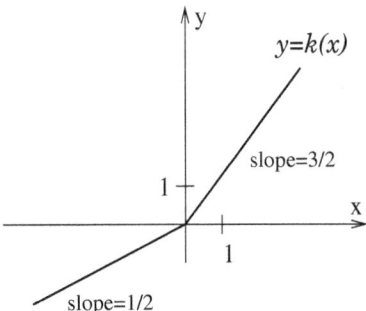

Since $y = h(x)$ can be obtained from $y = k(x)$ by shifting it 1 unit to the right and $\frac{1}{2}$ unit upward, $h(x) = k(x - 1) + \frac{1}{2}$.

Subtracting x from $k(x)$ gives $l(x) = k(x) - x$:

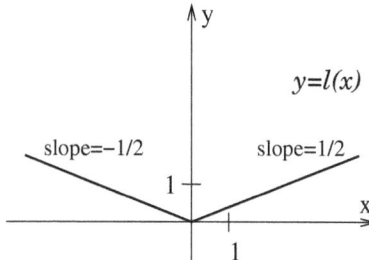

We see that $l(x) = \left|\dfrac{x}{2}\right|$.

The equation $l(x) = k(x) - x$ implies $k(x) = l(x) + x = \left|\dfrac{x}{2}\right| + x$.

Then $h(x) = k(x-1) + \dfrac{1}{2} = \left|\dfrac{x-1}{2}\right| + (x-1) + \dfrac{1}{2} = \left|\dfrac{x-1}{2}\right| + x - \dfrac{1}{2}$.

Further, $g(x) = |h(x)| = \left|\left|\dfrac{x-1}{2}\right| + x - \dfrac{1}{2}\right|$.

Finally, $g(x) = f(x) - \dfrac{x}{2}$ implies $f(x) = g(x) + \dfrac{x}{2} = \left|\left|\dfrac{x-1}{2}\right| + x - \dfrac{1}{2}\right| + \dfrac{x}{2}$.

11. *Solution 1.*

   Suppose the 4-tuple 0, 5, 0, 5 does occur. Then before it we must have the digit 0, and before that 5, and before that another 0... In fact, all the digits in our sequence must be 0's and 5's.

   Proof: Solving $a_n \equiv a_{n-4} + a_{n-3} + a_{n-2} + a_{n-1} \pmod{10}$ for a_{n-4} gives

   $$a_{n-4} \equiv a_n - a_{n-3} - a_{n-2} - a_{n-1} \pmod{10}.$$

   This implies $a_{n-4} \equiv a_n - a_{n-3} - a_{n-2} - a_{n-1} \pmod 5$.

   Thus if four consecutive digits are divisible by 5, then all the digits in the sequence are divisible by 5.

   But the starting sequence 2, 0, 0, 3 contains 2 and 3 which are not divisible by 5. Contradiction.

   Solution 2.

   If the 4-tuple 0, 5, 0, 5 does occur, look at *the very first time* it occurs. Then working backwards, determine a few digits before these 4:

   $$
   \begin{array}{ll}
   2,0,0,3,\ldots & \ldots,0,5,0,5,\ldots \\
   2,0,0,3,\ldots & \ldots,0,\underline{0,5,0,5},\ldots \\
   2,0,0,3,\ldots & \ldots,5,0,\underline{0,5,0,5},\ldots \\
   2,0,0,3,\ldots & \ldots,0,5,0,\underline{0,5,0,5},\ldots \\
   2,0,0,3,\ldots & \ldots,5,0,5,0,\underline{0,5,0,5},\ldots \\
   2,0,0,3,\ldots & \ldots,\underline{0,5,0,5},0,\underline{0,5,0,5},\ldots
   \end{array}
   $$

   We see that we have this 4-tuple in the sequence again, hence the one we started with was not the first occurrence. Contradiction.

13. We want to force our opponent to take the last counter. Thus we have to leave 1 counter on our last turn. To ensure that we'll be able to do that, we'll leave 6 counters on our next to last turn (then if our opponent takes 1, we take 4 and leave 1; if our

opponent takes 2, we take 3; if they take 3, we take 2; if they take 4, we take 1). On the turn before the next to last we'll leave 11... and so on. Thus we have to go first, take 1 counter and leave 26. Then no matter how our opponent plays we'll be able to leave 21, 16, 11, 6, 1.

15. On our last turn we want to leave one counter. Then our opponent will have to take it, and they will lose. Notice that no matter how our opponent plays, we can always play in such a way that the number of counters our opponent takes plus the number of counters we take is equal to 3 (namely, if they take 1, we can take 2; if they take 2, we can take 1). Thus on our next to last turn we'll leave 4 (then no matter how they play, we'll be able to leave 1). On the turn before that we want to leave 7. This means that we should let our opponent go first. Then, if they take x, we take $3 - x$, and leave 7. Then we leave 4, then we leave 1, and we win.

17. In order to win we want to take the last counter. Thus we will win if we get 1, 2, or 4 counters on our last turn. This means that if we leave 3 counters on the turn before that, our opponent will be able to take 1 or 2 counters leaving us with 2 or 1 respectively. In order to be able to leave 3, we want to get 4, 5, or 7 counters. Now notice that if we leave 6 on the turn before that, our opponent will be able to take 1, 2, or 4, leaving us with 5, 4, or 2 respectively, and all of these are winning positions for us. Thus we want to go first and take 4 counters on our first turn. This will leave 6. Then, as described above, we will win. (This is a winning strategy for the first player.)

19. We will work backwards. We want to take the last counter. How many counters should we leave on our next to last turn so that our opponent cannot take the last counter? We can't leave 1 because they will take it. We can't leave 2 either, since they can take 2. But we can leave 3. Then they can take either 1 or 2, that will leave 2 or 1 (respectively), and we will take them. So, we will leave 3 counters.

How many should we leave on the turn before that so that our opponent cannot take all or leave 3? We can't leave 4 (they may take all of them). We can't leave 5 (they may take 2 and leave 3). How about 6? Let's see what choices our opponent has then. If they take 1 and leave 5, that's good - we'll take 2 then. If they take 2 and leave 4, also good for us - we'll take 1. They can also take 4 and leave 2. That's good too, we will then take the last 2. So, it good for us to leave 6.

The turn before that: we don't want to give our opponent an opportunity to leave 0 or 3 or 6. Leaving 7 is bad (they may take 1 and leave 6), 8 is bad (they may take 2 and leave 6). How about 9? Our opponent may take 1 (and leave 8) or take 2 (and leave 7) or take 4 (and leave 5) or take 8 (and leave 1). In each case, we'll be able to leave 6 or 3 or 0. So, it is good for us to leave 9.

Now notice that the numbers 3, 6, 9 are multiples 3. Does this mean that leaving multiples of 3 is a winning strategy? Let's see... Suppose we leave a multiple of 3. Our opponent will take a power of 2. Since a power of 2 is not divisible by 3, they will leave a number not divisible by 3. Then we can take the remainder, and leave a multiple of 3 again. Thus we have to go first, take 2 counters, and leave 48 (or take 8 and leave 42, or take 32 and leave 18). Then, each time we'll be able to leave a multiple of 3. Thus sooner or later we'll leave 0, and we'll win.

21. If we want to win, we want to leave 1 match on our last turn so that our opponent loses. To be able to do this, we should get 2 matches on our last turn (if we get 3 or more, we won't be able to remove all but one). This means that we want to leave 3 the turn before that: our opponent can only remove 1 leaving us with 2. To be able to leave 3, we should get 4 or 5 or 6. Notice that if we leave 7 the turn before that, our opponent

will be able to take 1 or 2 or 3 leaving us with 6 or 5 or 4 respectively which is exactly what we want. We can leave 7 if we get any number from 8 to 14 inclusive. Thus if we leave 15 before that our opponent will be able to take 1-7 leaving us with 14-8. And so on... Notice that the numbers of matches we should leave on our turn (going backwards) are 1, 3, 7, 15. These all are one less than powers of 2. So it seems that the strategy is to always leave one less than a power of 2. Namely, we get a strategy for the first player: he/she should remove 45 matches leaving 255; the opponent will remove some number from 1 to 127 (inclusive), leaving some number between 254 and 128 (inclusive). The first player then will be able to remove the some number between 127 and 1 leaving 127 (which means that he/she removed no more than half); the opponent will remove some number between 1 and 63 leaving some number between 126 and 64. The first player should remove the required number between 63 and 1 to leave 63; the opponent will remove 1-31 leaving 62-32. The first player should remove the required number between 31 and 1 to leave 31; the opponent will remove 1-15 leaving 30-16. The first player should remove the required number between 15 and 1 to leave 15. Then, as described above, the first player should leave 7, 3, and, finally, 1, and he/she wins.

23. The derivative of a cubic polynomial is a quadratic polynomial. We want this quadratic polynomial to have integer roots. Instead of trying random coefficients a, b, c, and d, let's choose the roots of the quadratic polynomial (the derivative of f), and then find f:

Choose the roots, e.g. $r_1 = 3$ and $r_2 = 5$.

Then $(x-3)(x-5) = x^2 - 8x + 15$.

Now $f(x)$ can be any anti derivative of this polynomial, say, $\frac{1}{3}x^3 - 4x^2 + 15x - 3$. However, we want it to have integer coefficients, so let's multiply this function by 3:

$f(x) = x^3 - 12x^2 + 45x - 9$. Then $f'(x) = 3x^2 - 24x + 45 = 3(x^2 - 8x + 15) = 3(x-3)(x-5)$ has integer roots.

Here is another choice of roots and the constant d:

$r_1 = -3$, $r_2 = 4$, $(x+3)(x-4) = x^2 - x - 12$, an antiderivative is $\frac{1}{3}x^3 - \frac{1}{2}x^2 - 12x - \frac{1}{6}$, multiply by 6: $f(x) = 2x^3 - 3x^2 - 72x - 1$. Then $f'(x) = 6x^2 - 6x - 72 = 6(x^2 - x - 12) = 6(x+3)(x-4)$ has integer roots.

25. Start with a matrix in reduced echelon form with integer entries, and perform a few operations (i.e. work backwards in the reducing algorithm) to modify some (or all) coefficients.

For example:

$$\begin{bmatrix} 1 & 0 & 3 \\ 0 & 1 & -2 \\ 0 & 0 & 0 \end{bmatrix} \leftarrow \begin{bmatrix} 1 & 0 & 3 \\ 0 & 1 & -2 \\ 0 & -2 & 4 \end{bmatrix} \leftarrow \begin{bmatrix} 1 & 1 & 1 \\ 0 & 3 & 6 \\ 0 & -2 & 4 \end{bmatrix} \leftarrow \begin{bmatrix} 1 & 1 & 1 \\ -2 & 1 & 4 \\ 0 & -2 & 4 \end{bmatrix} \leftarrow$$

$$\begin{bmatrix} 1 & 1 & 1 \\ -2 & 1 & 4 \\ 3 & 1 & 7 \end{bmatrix} \leftarrow \begin{bmatrix} 4 & 4 & 4 \\ -2 & 1 & 4 \\ 3 & 1 & 7 \end{bmatrix}$$

27. Based on the amount we see that coins of some value must be present. For example, at least two pennies must be used to make 57 cents. Instead of adding the values of coins, we will subtract the value of known coins from the initial amount and keep track of the number of other coins.

So, since at least two pennies must be present, the value of the other 5 coins is 55 cents. These 5 coins cannot all be less than a quarter, because 5 coins whose value is at most 10 cents would not add up to 55, and no coins can be larger than a quarter since the next available value is 50 cents, and that would leave 5 cents for 4 coins which is impossible. So at least one quarter must be present. Thus the value of the remaining 4 coins is $55 - 25 = 30$. Now, no pennies or coins larger than 10 cents can be used, and four dimes would be too much, so at least one nickel is present. The value of the 3 remaining coins is 25, so again at least one nickel must be present, and the value of the remaining 2 coins is 20. The only possibility here is two dimes.

Thus I have two pennies, two nickels, two dimes, and one quarter.

17.15 Calculus

1. Since $|x + 2| = \begin{cases} x + 2 & \text{if } x + 2 \geq 0, \text{ i.e. } x \geq -2 \\ -(x + 2) & \text{if } x + 2 \leq 0, \text{ i.e. } x \leq -2 \end{cases}$, we have:

$$\int_{-4}^{2} |x + 2| dx = \int_{-4}^{-2} |x + 2| dx + \int_{-2}^{2} |x + 2| dx = -\int_{-4}^{-2} (x + 2) dx + \int_{-2}^{2} (x + 2) dx =$$
$$-\left(\frac{x^2}{2} + 2x\right)\Big|_{-4}^{-2} + \left(\frac{x^2}{2} + 2x\right)\Big|_{-2}^{2} = -(2 - 4) + (8 - 8) + (2 + 4) - (2 - 4) = 10$$

Note: another way to do this problem is to interpret the integral in terms of areas.

3. Let the given line be tangent to the parabola at the point $(a, a - 1)$. Then first, the parabola passes through $(a, a - 1)$, thus

$$a - 1 = ca^2.$$

Second, the line and the parabola have the same slope at this point:

$$1 = 2ca.$$

From the second equation we have $c = \dfrac{1}{2a}$. Substitute this for c in the first equation:

$$a - 1 = \frac{a^2}{2a} \quad \Rightarrow \quad a - 1 = \frac{a}{2} \quad \Rightarrow \quad 2a - 2 = a \quad \Rightarrow \quad a = 2 \quad \Rightarrow \quad c = \frac{1}{4}.$$

5. We need the polynomial to pass through the given points, and have slope (which is $p'(x) = 3ax^2 + 2bx + c$) equal to 0 at both points.
The value at 0: $d = 1$.
The value at 1: $a + b + c + d = 0$.
The slope at 0: $c = 0$.
The slope at 1: $3a + 2b + c = 0$.
Since $d = 1$ and $c = 0$, the second and fourth equations become $a + b = -1$ and $3a + 2b = 0$. Then $b = -\frac{3}{2}a$, and $a - \frac{3}{2}a = -1$. This gives $a = 2$. Then $b = -3$.
So $p(x) = 2x^3 - 3x^2 + 1$.

7. To find the intersection points of the line $y = ax$ and the parabola $y = x^2$, solve $ax = x^2$.

The roots are $x = 0$ and $x = a$, thus the intersection points are $(0, 0)$ and (a, a^2).

If $a > 0$, the area is $\displaystyle\int_0^a (ax - x^2) dx = \left(a\frac{x^2}{2} - \frac{x^3}{3}\right)\Big|_0^a = \frac{a^3}{2} - \frac{a^3}{3} = \frac{a^3}{6}.$

We want the area to be equal to 1, so $\dfrac{a^3}{6} = 1 \quad \Rightarrow \quad a^3 = 6 \quad \Rightarrow \quad a = \sqrt[3]{6}.$

If $a < 0$, then the area is $\displaystyle\int_a^0 (ax - x^2)\,dx = -\dfrac{a^3}{6} \quad \Rightarrow \quad a = -\sqrt[3]{6}.$

9. $\displaystyle\sum_{n=0}^{\infty} \frac{1}{2^{2n+1}} = \frac{1}{2} + \frac{1}{2^3} + \frac{1}{2^5} + \frac{1}{2^7} + \ldots = \frac{1}{2}\left(1 + \frac{1}{2^2} + \frac{1}{2^4} + \frac{1}{2^6} + \ldots\right)$

$= \dfrac{1}{2}\left(1 + \dfrac{1}{4} + \dfrac{1}{4^2} + \dfrac{1}{4^3} + \ldots\right) = \dfrac{1}{2} \cdot \dfrac{1}{1 - \frac{1}{4}} = \dfrac{2}{3}$

11. Draw a picture so that you see what's going on. Let the slope of such a tangent line be m, then its equation is $y = mx$. Let (a, ma) be the touching point. Since this point lies on the parabola, $ma = a^2 + 2$. The slope of the parabola at the touching point must be m, therefore $2a = m$. Substituting this into the first equation gives $2a^2 = a^2 + 2$. Then $a = \pm\sqrt{2}$, and $m = \pm 2\sqrt{2}$. Thus the equations of the tangent lines are $y = 2\sqrt{2}x$ and $y = -2\sqrt{2}x$.

13. First find the partial fraction decomposition, i.e. A and B such that

$$\frac{1}{x^2 + x} = \frac{A}{x} + \frac{B}{x+1}.$$

Multiply both sides by $x^2 + x = x(x + 1)$:

$1 = A(x + 1) + Bx$

$1 = (A + B)x + A \qquad \Rightarrow \qquad A + B = 0 \text{ and } A = 1, \text{ then } B = -1.$

Thus $f(x) = \dfrac{1}{x^2 + x} = \dfrac{1}{x} - \dfrac{1}{x+1} = x^{-1} - (x+1)^{-1}.$

$f'(x) = -x^{-2} + (x+1)^{-2}$

$f''(x) = 2x^{-3} - 2(x+1)^{-3}$

$f'''(x) = -2 \cdot 3x^{-4} + 2 \cdot 3(x+1)^{-4}$

$\ldots$

$f^{(n)}(x) = (-1)^n n! x^{-n-1} - (-1)^n n! (x+1)^{-n-1}$

Note: this formula can be proved by Mathematical Induction.

15. Since A and B are given, the length of AB is given. Now, to maximize the area of $\triangle ABC$, we have to maximize the height h_c. To do this, the point C must lie on the tangent line parallel to the given line. Thus the slope of the parabola at C must be equal to m. Then the x-coordinate of C is $\dfrac{m}{2}$ (since the slope is $2x$). The y-coordinate of C is then $\dfrac{m^2}{4}$. It can be verified that this point is always between the intersection points A and B.

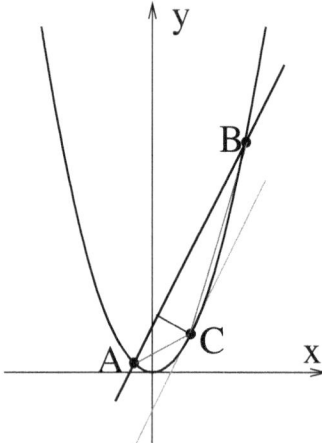

17. Consider the graphs of $f(x) = x^2 + ax + 1$ and $g(x) = \cos x$. Both graphs pass through the point $(0, 1)$. The graph of $g(x) = \cos x$ has slope 0 at that point. If the slope of $f(x) = x^2 + ax + 1$ at $(0, 1)$ is positive, then for some small negative values of x we will have $x^2 + ax + 1 < \cos x$. If the slope of $f(x) = x^2 + ax + 1$ at $(0, 1)$ is negative, then for some small positive values of x we will have $x^2 + ax + 1 < \cos x$. The only case in which $x^2 + ax + 1 \geq \cos x$ for all real x is when the slope of $f(x) = x^2 + ax + 1$ at $(0, 1)$ is 0. The derivative of $f(x)$ is $f'(x) = 2x + a$, thus the slope at $(0, 1)$ is $f'(0) = a$. Therefore $a = 0$ is the only such value of a.

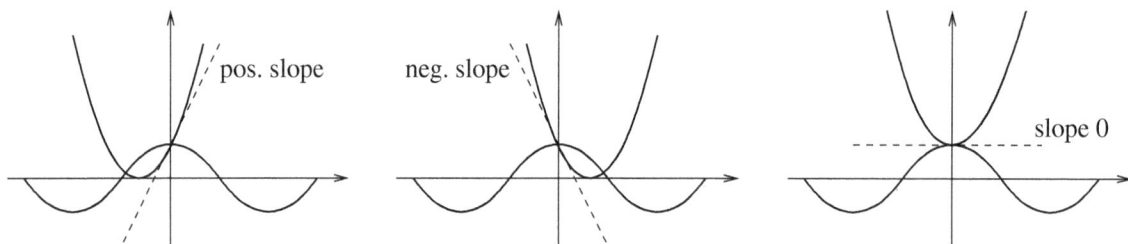

19. Let point A (with positive x-coordinate) where the circle touches the parabola be (a, a^2), and the center B of the circle be $(0, b)$. Then the distance between these points is 1, thus

$$a^2 + (b - a^2)^2 = 1.$$

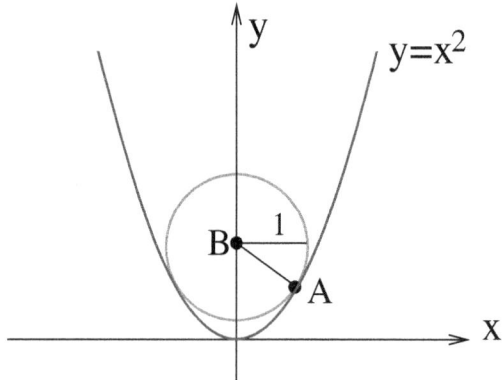

The slope of the parabola at the point (a, a^2) is $2a$ (the derivative of x^2 at $x = a$), then the slope of AB is $-\dfrac{1}{2a}$ (since AB and the parabola are orthogonal at (a, a^2)). Thus we have

$$\frac{b - a^2}{0 - a} = -\frac{1}{2a}.$$

The last equation gives $b - a^2 = \dfrac{1}{2}$, then from the first equation we have

$a^2 + \dfrac{1}{4} = 1 \quad \Rightarrow \quad a^2 = \dfrac{3}{4}$. Then $b = a^2 + \dfrac{1}{2} = \dfrac{3}{4} + \dfrac{1}{2} = \dfrac{5}{4}$. So the center of the circle is at $(\dfrac{5}{4}, 0)$.

21. Our region consists of 4 parts of equal area.

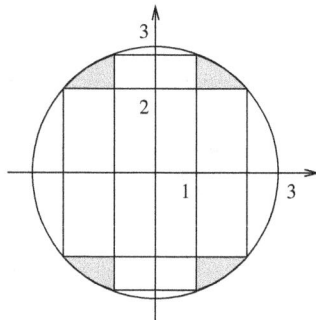

The total area of our region is 4 times the area of each part.

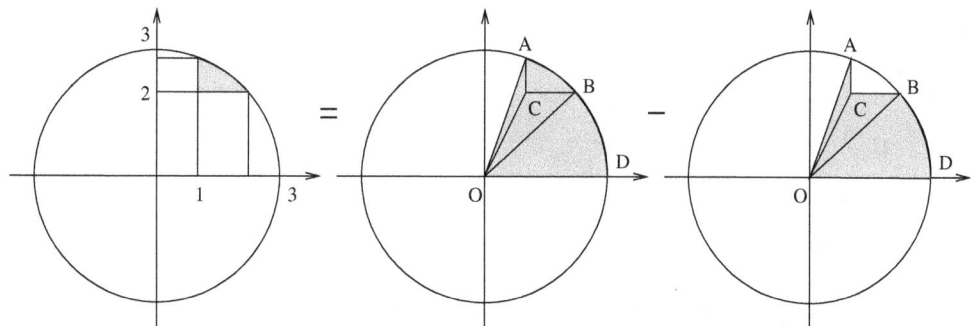

The area of each part is the area of sector OAD minus the area of sector OBD minus the area of triangle OBC minus the area of triangle OAC.

A has coordinates $(1, \sqrt{8})$, thus $\text{Area}_{OAD} = \dfrac{9 \arccos(1/3)}{2}$.

B has coordinates $(\sqrt{5}, 2)$, thus $\text{Area}_{OBD} = \dfrac{9 \arcsin(2/3)}{2}$.

OBC has base $BC = \sqrt{5} - 1$ and height $h_{BC} = 2$, thus $\text{Area}_{OBC} = \dfrac{2(\sqrt{5} - 1)}{2}$.

OAC has base $AC = \sqrt{8} - 2$ and height $h_{AC} = 1$, thus $\text{Area}_{OAC} = \dfrac{\sqrt{8} - 2}{2}$.

Then $\text{Area}_{ABC} = \dfrac{9 \arccos(1/3) - 9 \arcsin(2/3) - 2(\sqrt{5} - 1) - (\sqrt{8} - 2)}{2}$

$$= \frac{9\arccos(1/3) - 9\arcsin(2/3) - 2\sqrt{5} - \sqrt{8} + 4}{2}$$

The total area is then $2(9\arccos(1/3) - 9\arcsin(2/3) - 2\sqrt{5} - \sqrt{8} + 4)$

23. Let the right intersection point have coordinates (a, c). Then $c = 8a - 27a^3$.

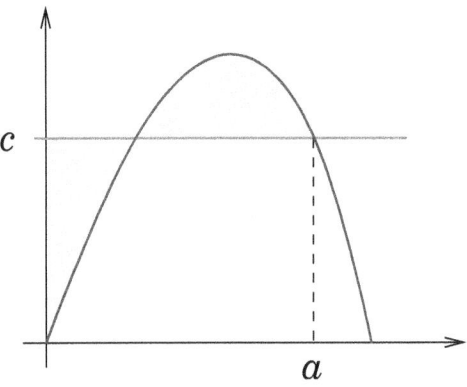

If the areas of the shaded regions are equal then the area of the region under the given cubic curve from $x = 0$ to $x = a$ is equal to the area of the rectangle with width a and height $c = 8a - 27a^3$. Thus we have

$$\int_0^a 8x - 27x^3 = a(8a - 27a^3)$$

$$\left(4x^2 - \frac{27}{4}x^4\right)\Big|_0^a = 8a^2 - 27a^4$$

$$4a^2 - \frac{27}{4}a^4 = 8a^2 - 27a^4$$

$$\frac{81}{4}a^4 = 4a^2$$

$$81a^4 = 16a^2$$

$$81a^2 = 16 \text{ (since } a \neq 0)$$

$$a = \sqrt{\frac{16}{81}} = \frac{4}{9}$$

Then $c = 8a - 27a^3 = \dfrac{32}{9} - \dfrac{27 \cdot 4^3}{9^3} = \dfrac{32}{9} - \dfrac{64}{27} = \dfrac{96 - 64}{27} = \dfrac{32}{27}$.

25. Assume that a, b, and c are all positive.

Case 1. At least 2 of a, b, and c are equal. Without loss of generality we can assume that $a = b$. Then the intersection of the xy-plane and the ellipsoid is a circle whose equation is $\dfrac{x^2}{a^2} + \dfrac{y^2}{b^2} = 1$, $z = 0$.

Case 2. The numbers a, b, and c are all distinct. Without loss of generality we can assume that $a > b > c$. The intersection of the xy-plane and the ellipsoid is an ellipse whose equation is $\dfrac{x^2}{a^2} + \dfrac{y^2}{b^2} = 1$, $z = 0$. The intersection of the zy-plane and the ellipsoid is an ellipse whose equation is $\dfrac{z^2}{c^2} + \dfrac{y^2}{b^2} = 1$, $x = 0$. Now rotate the xy-plane until it coincides with the zy-plane so that the y-axis is always in it. The intersection of this rotating plane with the ellipsoid is always an ellipse. Moreover, one of its axes is always b, and the other one is changing continuously from $a > b$ to $c < b$. Because of continuity, at some point the changing axis is equal to b.

27. If the function $a_1 \cos x + a_2 \cos(2x) + \ldots + a_{30} \cos(30x)$ takes on only positive values, then its integral over any interval must be positive (because it is the area of the region under the graph of the function). However,

$$\int_0^{2\pi} (a_1 \cos x + a_2 \cos(2x) + \ldots + a_{30} \cos(30x))dx =$$

$$\left(a_1 \sin x + \frac{a_2}{2} \sin(2x) + \ldots + \frac{a_{30}}{30} \sin(30x)\right)\Big|_0^{2\pi} = 0.$$

29. At $x = 0$ we have $\dfrac{x}{x^2 + 1} = \arctan x = x$.

Let's compare the derivatives of these functions for $x > 0$. The derivatives are:

$$\left(\frac{x}{x^2 + 1}\right)' = \frac{1 - x^2}{(x^2 + 1)^2},$$

$$(\arctan x)' = \frac{1}{x^2 + 1} = \frac{1 + x^2}{(x^2 + 1)^2},$$

$$(x)' = 1 = \frac{1 + x^2}{x^2 + 1}.$$

We see that for all $x > 0$, $\left(\dfrac{x}{x^2 + 1}\right)' < (\arctan x)' < (x)'$, therefore the curve $y = \dfrac{x}{x^2 + 1}$ lies below the curve $y = \arctan x$ which lies below the line $y = x$.

31. The 4-dimensional volume of a 4-dimensional ball is proportional to the fourth power of its radius. Suppose $V = cr^4$ where c is a constant. Then the 3-dimensional volume of the boundary of this ball is $S = V' = 4cr^3$. Therefore $\dfrac{V}{S} = \dfrac{cr^4}{4cr^3} = \dfrac{r}{4}$.

33. $\displaystyle\int_0^1 (\sqrt[3]{1 - x^7} - \sqrt[7]{1 - x^3})dx = \int_0^1 \sqrt[3]{1 - x^7}dx - \int_0^1 \sqrt[7]{1 - x^3}dx.$

The integral $\displaystyle\int_0^1 \sqrt[3]{1 - x^7}dx$ is equal to the area of the region bounded by $y = \sqrt[3]{1 - x^7}$, the x-axis, and the y-axis.

The integral $\displaystyle\int_0^1 \sqrt[7]{1 - x^3}dx$ is equal to the area of the region bounded by $y = \sqrt[7]{1 - x^3}$, the x-axis, and the y-axis.

Equations $y = \sqrt[3]{1 - x^7}$ and $y = \sqrt[7]{1 - x^3}$ can be rewritten as $x^7 + y^3 = 1$ and $x^3 + y^7 = 1$ respectively. It is easy to see that both curves pass through $(1, 0)$ and through $(0, 1)$, and these two curves are symmetric about the line $y = x$. Thus the areas of the two regions described above are equal, therefore the difference of the integrals $\displaystyle\int_0^1 \sqrt[3]{1 - x^7}dx$ and $\displaystyle\int_0^1 \sqrt[7]{1 - x^3}dx$ is 0. Thus $\displaystyle\int_0^1 (\sqrt[3]{1 - x^7} - \sqrt[7]{1 - x^3})dx = 0.$

35. Since $f(0) = 0$ and $\sin(0) = 0$, $|f(x)| < |\sin(x)|$ for all x, and the slope of $y = \sin(x)$ at $(0, 0)$ is 1, we have $|f'(0)| < 1$.

Since $|f'(0)| = |a_1 + 2a_2 + \ldots + na_n|$, the required inequality follows.

37. Let the curve be given by $y = f(x)$. Since it passes through $(3, 2)$, $f(3) = 2$.

At a point $P(a, f(a))$, the tangent line has slope $f'(a)$ and equation $y - f(a) = f'(a)(x - a)$. Its x-intercept is $\left(0, -\dfrac{f(a)}{f'(a)} + a\right)$. The part of the tangent line that lies in the first quadrant is bisected by P iff $2a = -\dfrac{f(a)}{f'(a)} + a$. Thus $af'(a) = -f(a)$. Since this must be true for every point on the curve in the first quadrant, we have the differential equation $xf'(x) = -f(x)$. Any function of the form $f(x) = \dfrac{c}{x}$ is a solution of this equation. Using the condition $f(3) = 2$, we find $c = 6$. So $f(x) = \dfrac{6}{x}$ satisfies the required condition.

39. Since $\arcsin(x)$ (defined on $[-1, 1]$) is the inverse function of $\sin(x)$ on $\left[-\dfrac{\pi}{2}, \dfrac{\pi}{2}\right]$, their graphs are symmetric about the line $y = x$, thus the value of $\displaystyle\int_0^1 \arcsin(x)dx$, the area of the region under the graph of $\arcsin(x)$ on $[0, 1]$, is equal to the area of the region between $y = \sin(x)$ and $y = 1$ from $x = 0$ to $x = \dfrac{\pi}{2}$. (Draw the graphs to see this!) The latter area can be calculated by $\displaystyle\int_0^{\frac{\pi}{2}} (1 - \sin(x))dx = (x + \cos(x)) \Big|_0^{\frac{\pi}{2}} = \dfrac{\pi}{2} - 1$.

17.16   Various problems

1. *Proof 1:* In a reentrant knight's tour black and white squares must alternate. But a 5×5 chessboard has 13 squares of one color and 12 squares of the other color, so it is not possible to have a cycle in which the colors alternate.

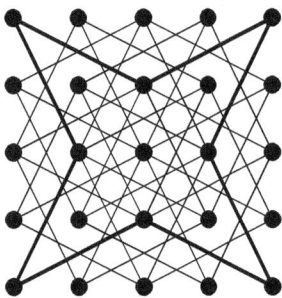

Proof 2: Draw a graph representing legal moves of a knight. Look at the corner vertices. They all have degree 2, thus, in order to visit the corner vertices, we must use both edges at each corner vertex. Those 8 edges form a cycle. It is not possible to add more edges to this cycle, but the cycle misses many points. Therefore there is no Hamilton cycle, and thus there is no reentrant tour.

3. *Sketch:* Calculate first few terms, and notice that $a_n = \dfrac{1}{n!}$. Prove this formula by induction. Then $\displaystyle\sum_{n=0}^{\infty} a_n = \sum_{n=0}^{\infty} \dfrac{1}{n!} = e$.

5. *Solution 1.* Consider positive and negative values of x separately:

Case I. $x \geq 0$, then $|x| = x$, and the inequality becomes $|6 - 2x| + x \leq 3$.

Case Ia. If $6 - 2x \geq 0$, or $x \leq 3$, then $|6 - 2x| = 6 - 2x$, and we have $6 - 2x + x \leq 3$. This gives $x \geq 3$. Together with the condition $x \leq 3$, we get one root $x = 3$. This root satisfies the condition $x \geq 0$.

Case Ib. If $6 - 2x < 0$, or $x > 3$, then $|6 - 2x| = -6 + 2x$, and we have $-6 + 2x + x \le 3$. This gives $x \le 3$ which contradicts the condition $x > 3$. Thus we have no roots in this case.

Case II. $x < 0$, then $|x| = -x$, and the inequality becomes $|6| + x \le 3$, or $6 + x \le 3$ since $|6| = 6$. Equivalently, $x \le -3$. All the values of $x \le -3$ satisfy the condition $x < 0$.

Answer: $x = 3$ and $x \le -3$. In the interval notation, $(-\infty, -3] \cup \{3\}$.

Solution 2. Draw the graph of $f(x) = |6 - |x| - x| + x$.

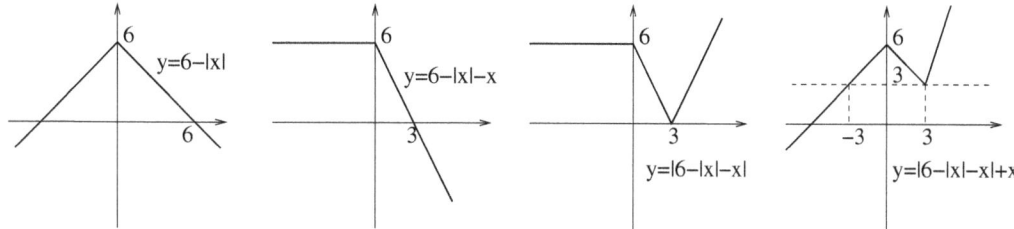

We see that $f(x) \le 3$ when $x = 3$ and when $x \le 3$.

7. Below are some examples (but there are many others).

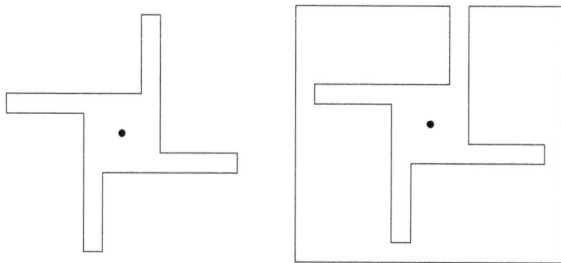

9. Recall problems 19 and 20 in chapter 5. Generalize to a $(n+1) \times \left(\dfrac{(n+1)n^2}{2} + 1 \right)$ board colored with n colors. Then dirive the desired result.

11. First notice that years 2008, 2012, 2016, 2020, and 2024 are leap years. The other 15 years from 2006 to 2025 (including these) are not. Since non-leap years contain 365 days and leap years contain 366 days, the number of days that pass between December 25, 2005 and December 25, 2025 is $15 \cdot 365 + 5 \cdot 366 \equiv 1 \cdot 1 + 5 \cdot 2 \equiv 11 \equiv 4 \pmod 7$, therefore December 25, 2025 is 4 days after a Sunday, i.e. is a Thursday.

13. Suppose that the 9 bags contain different numbers of coins, and the total number of coins is 40. Let $a_1 < a_2 < \ldots < a_9$ be the numbers of coins in the 9 bags. Then $a_1 \ge 1$, $a_2 \ge 2$, ..., $a_9 \ge 9$, and $40 = a_1 + a_2 + \ldots + a_9 \ge 1 + 2 + \ldots + 9 = 45$. Since $40 < 45$, we get a contradiction, thus it is not possible for all bags to contain different numbers of coins.

www.ingramcontent.com/pod-product-compliance
Lightning Source LLC
Chambersburg PA
CBHW081128170526
45165CB00008B/2591